T0237450

SpringerBriefs in Physics

SpringerBriefs in Physics are a series of slim high-quality publications encompassing the entire spectrum of physics. Manuscripts for SpringerBriefs in Physics will be evaluated by Springer and by members of the Editorial Board. Proposals and other communication should be sent to your Publishing Editors at Springer.

Featuring compact volumes of 50 to 125 pages (approximately 20,000–45,000 words), Briefs are shorter than a conventional book but longer than a journal article. Thus, Briefs serve as timely, concise tools for students, researchers, and professionals.

Typical texts for publication might include:

- A snapshot review of the current state of a hot or emerging field
- A concise introduction to core concepts that students must understand in order to make independent contributions
- An extended research report giving more details and discussion than is possible in a conventional journal article
- A manual describing underlying principles and best practices for an experimental technique
- An essay exploring new ideas within physics, related philosophical issues, or broader topics such as science and society

Briefs allow authors to present their ideas and readers to absorb them with minimal time investment. Briefs will be published as part of Springer's eBook collection, with millions of users worldwide. In addition, they will be available, just like other books, for individual print and electronic purchase. Briefs are characterized by fast, global electronic dissemination, straightforward publishing agreements, easy-to-use manuscript preparation and formatting guidelines, and expedited production schedules. We aim for publication 8–12 weeks after acceptance.

More information about this series at http://www.springer.com/series/8902

Wolfgang Bacsa · Revathi Bacsa ·
Tim Myers

Optics Near Surfaces
and at the Nanometer Scale

 Springer

Wolfgang Bacsa
Physique
CEMES CNRS et Université de Toulouse
Toulouse, France

Revathi Bacsa
RR Bacsa Scientific
Castanet-Tolosan, France

Tim Myers
Industrial Mathematics
Centre de Recerca Matemàtica
Bellaterra (Barcelona), Spain

ISSN 2191-5423   ISSN 2191-5431   (electronic)
SpringerBriefs in Physics
ISBN 978-3-030-58982-0   ISBN 978-3-030-58983-7   (eBook)
https://doi.org/10.1007/978-3-030-58983-7

This Springer imprint is published by the registered company Springer Nature Switzerland AG
The registered company address is: Gewerbestrasse 11, 6330 Cham, Switzerland

To Anna and Tomás for not always disturbing me while writing this book during lockdown (well at least one of them didn't)

—*T.M.*

Preface

Optical methods are powerful tools to observe and understand the world we live in. Today the availability of advanced light sources and high-precision detectors provides us with new improved optical microscopic and spectroscopic techniques to understand the nanoworld in a noninvasive way. Thus, controlling matter at increasingly smaller scales goes hand in hand with extending the application domain of optical observation tools. Geometrical optics as taught at high school and university levels is founded on ray optics, even while acknowledging that at very small scales or in the nano-regime comprising ultra-small particles, molecules or atomic layers, the wave nature of light needs to be considered.

The wave nature of light was first invoked by Wiener in 1890 who used interference near surfaces to demonstrate this concept. These first experiments could effectively explain the colours of thin films and led to the work of Lippman who discovered a method to reproduce colours in photography based on the interference phenomenon, for which he received the Nobel prize in 1908. Using advanced tools, today we have the possibility to explore optical fields in much greater detail which is crucial for further advancement in several fields such as plasmon optics or meta materials.

In this book we will discuss wave optics near surfaces. We also describe how a closer look at interference fringes in the proximity of surfaces provides the opportunity to enhance optical signals from nanoparticles, molecules and atomic layers or to extend optical holography to the sub-micrometer range. In contrast to near-field optics where fields in the ultimate proximity of the material are recorded ($< \lambda/10$) the details of which are described in many works, this book focuses on optical fields at larger distances, typically at distances as large as the size of the image, this intermediate range being distinctly different from the far field range. In short, near-field optics explores the quasi-electrostatic longitudinal field range, whereas in the intermediate field, the propagating transverse field component is explored.

Chapter 1 gives an introduction to electromagnetic waves in vacuum and in materials. Chapter 2 describes optical surface standing waves and how interference substrates can be used to increase the optical response of molecules and nanoparticles, after which a short review on how this knowledge can be exploited in various applications in spectroscopy and in imaging atomic layers is given. Chapter 3 describes the formation of interference fringes due to the presence of a single point scatterer on a surface. The analytical description allows for the extraction of parameters from the images of the interference fringes and the reconstruction of interferometric images recorded in the intermediate field. Chapter 4 explores the consequences of the finite size of the optical focal point on optical spectroscopy and Chap. 5 shows how the index of refraction can be derived from a linear superposition of point dipoles.

This book is aimed at better understanding optical interference and diffraction near surfaces and we hope that it will be of use to both researchers and graduate students looking for an introduction to optics at small scales. We have tried to make the mathematical description sufficiently comprehensive so that it is accessible to readers outside the field of physics.

Toulouse, Barcelona   Wolfgang Bacsa
July 2020   Revathi Bacsa
                                                                                 Tim Myers

Acknowledgements

The authors and in particular T. M. would like to acknowledge support from the Agence Nationale de Recherche, through the EUR grant NanoX n° ANR-17-EURE-0009 in the framework of the "Programme des Investissements d'Avenir". During the early stages of this collaboration funding was received through COST Action, TD1409, Mathematics for Industry Network and also the Spanish Ministerio de Ciencia e Innovación Grant No. MTM2017-82317-P.

Contents

Chapter 1
Introduction: Wave Optics Near Surfaces

1.1 Waves and Wave Scattering

The propagation of light is described by the wave equation. When propagating through a medium or, in other words, in the presence of matter, the speed of propagation of a wave is reduced, which is accounted for by the index of refraction. Since light is an electromagnetic wave, the index of refraction of the medium is related to the electric and magnetic susceptibility of the material. One may ask then, what is the origin of this reduction in the speed of light propagation? In geometrical optics, the index of refraction is described as changing abruptly at the interface between the medium and air while considering this interface to be perfectly flat. Optical rays are refracted at the interface according to Snell's law, which can be derived by applying Maxwell's equations across the interface [1]. While the behavior of electromagnetic waves is perfectly well described on the macroscopic scale, a physical understanding at smaller scales requires that we consider the finite size of the wavelength as well as the sizes of individual scatterers. Figure 1.1 illustrates the well-known process of optical reflection at the optical wavelength scale. It shows an incident and a reflected beam at a flat interface. The finite diameter of the incident beam causes the incident and reflected beams to overlap. Coherence in the monochromatic incident beam leads to interference which means that the overlapping beams form optical standing waves oriented parallel to the interface. For dielectric materials, it turns out that the interference is destructive at the surface and so the field in close proximity to the surface is actually quite small.

In the visible spectral range the wavelength (400–700 nm) is large compared to the size of atoms (0.1 nm) and so a uniform optical field may be assumed when working at the scale of atoms or small molecules. Richard Feynman [2] described the interaction of light with atoms as a process where the scattering of light occurs through absorption and emission of photons. Victor Weisskopf [3], one of the earliest physicists to examine light scattering at the atomic scale, considered atoms as electron oscillators driven by an incident electromagnetic wave. The scattered wave therefore

W. Bacsa et al., *Optics Near Surfaces and at the Nanometer Scale*, SpringerBriefs in Physics, https://doi.org/10.1007/978-3-030-58983-7_1

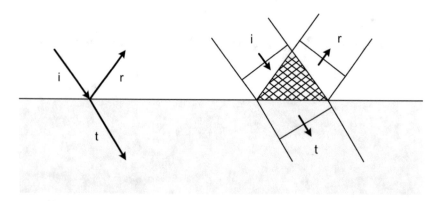

Fig. 1.1 At a perfectly flat interface between a medium and air, the incident beams splits into refracted and reflected beams. In the second figure we see that the finite width of the beam leads to a region of overlap of the incident and reflected beams (hatched triangle)

depends on the resonance frequency of the oscillator which then determines both the amplitude and the phase of the scattered wave. This means that the scattering of light by atoms leads necessarily to a phase shift of the scattered wave. The phase shift far below resonance is $-\pi/2$ while far above resonance it is $+\pi/2$. Put in another way, the incident wave polarizes the atom inducing a phase shifted dipole wave which propagates in all directions except in the direction of polarization. In the situation where there are many atoms, the many scattered waves are superimposed. Furthermore, since light travels at a finite speed, when emanating from scatterers separated in space the scattered waves will be phase shifted. Overall, this leads to an accumulated phase shift which results in a reduction of the speed of propagation of light in matter. Finally, the reflected and transmitted waves are formed through the interference of all the scattered waves with the incident wave in the forward and backward directions [4].

The wave behaviour discussed above may be described by Maxwell's equations. In the following section we will introduce these equations and show how they can be used for the wave description of light. The equations are essential for much of the work in this book and we will return to them in subsequent chapters.

1.2 Maxwell's Equations and Electromagnetic Waves

Electric fields are generated by charges and electric currents generate magnetic fields. On a macroscopic scale the behavior of electric and magnetic fields is described by Maxwell's equations which define how longitudinal and transverse variations in the field relate and how they are connected to their sources. Maxwell's equations are a set of partial differential equations which, when coupled with the Lorentz force

law, form the basis of classical electromagnetism and classical wave optics. The
Maxwell's equations in free space are

$$\nabla \cdot \mathbf{E} = \frac{\rho}{\epsilon_0} \tag{1.1}$$

$$\nabla \cdot \mathbf{B} = 0 \tag{1.2}$$

$$\nabla \times \mathbf{E} = -\frac{\partial \mathbf{B}}{\partial t} \tag{1.3}$$

$$\nabla \times \mathbf{B} = \mu_0 \left(\mathbf{J} + \epsilon_0 \frac{\partial \mathbf{E}}{\partial t} \right), \tag{1.4}$$

where $\mathbf{E}$ is the electric field, $\mathbf{B}$ the magnetic field (both are vectors depending on position and time), $\mathbf{J}$ is the current density, ρ is the electric charge density (charge per unit
volume) and ϵ_0, μ_0 are the permittivity and permeability of free space respectively.

Now we focus on the electric field and assume that the region of interest has no
source terms, that is, the charge and current densities are zero (or at least negligible);
this means that the source terms are outside of the considered domain. Taking the
curl of Eq. (1.3) and using Eq. (1.4) to eliminate the magnetic field results in

$$\nabla \times (\nabla \times \mathbf{E}) = -\frac{\partial}{\partial t} \nabla \times \mathbf{B} = -\mu_0 \epsilon_0 \frac{\partial^2 \mathbf{E}}{\partial t^2}. \tag{1.5}$$

Using the vector identity

$$\nabla \times (\nabla \times \mathbf{E}) = \nabla (\nabla \cdot \mathbf{E}) - \nabla^2 \mathbf{E} \tag{1.6}$$

and noting that $\nabla \cdot \mathbf{E} = 0$ we find

$$\nabla^2 \mathbf{E} = \mu_0 \epsilon_0 \frac{\partial^2 \mathbf{E}}{\partial t^2}. \tag{1.7}$$

We could just as easily have taken the curl of (1.4) and ended up with exactly the
same equation but with $\mathbf{E}$ replaced by $\mathbf{B}$. Equation (1.7) is a standard wave equation
which then indicates that the electric field (and magnetic field) travels as a wave.
Further, it turns out that in a vacuum $\sqrt{\mu_0 \epsilon_0} = 1/c$, where c is the speed of light.

The wave equation is frequently encountered in the fields of fluid mechanics,
acoustics and electromagnetism. This means we have a wealth of experience and
previous experiments to draw from. A particularly useful feature of the wave equation
is that it is linear. This means that if we have more than one solution, say $\mathbf{E}_1$, $\mathbf{E}_2$ then
the sum $\mathbf{E} = \mathbf{E}_1 + \mathbf{E}_2$ is also a solution. We will discuss this *superposition* of waves
in Chap. 2. The standard solution to the wave equation is a vector of the form

$$\mathbf{E} = \mathbf{E_m} \cos(\mathbf{k.r} - \omega t + \psi) \tag{1.8}$$

where $\mathbf{E_m}$ represents the maximum amplitude and direction of the oscillations, $\mathbf{k}$ is the wave vector which specifies the direction of travel, $\omega = c|\mathbf{k}|$ is the angular frequency and ψ is the phase angle. Since cosine is an even function, $\cos(x) = \cos(-x)$, the form $\cos(\omega t - \mathbf{k.r} + \psi)$ is equivalent and is employed in many texts (then ψ is simply the negative of that in (1.8)). For simplicity in mathematical calculations the wave is often written in its complex form

$$\mathbf{E} = \Re\left[\mathbf{E_m} e^{i(\mathbf{k.r} - \omega t + \psi)}\right] \tag{1.9}$$

where $\Re$ indicates the real part, which corresponds to the physical solution. If we substitute this form into the wave equation, (1.7), we find

$$\nabla^2 \mathbf{E} = -|\mathbf{k}|^2 \mathbf{E} , \qquad \frac{\partial^2 \mathbf{E}}{\partial t^2} = -\omega^2 \mathbf{E} , \tag{1.10}$$

and hence

$$|\mathbf{k}|^2 \mathbf{E} = \mu_0 \epsilon_0 \omega^2 \mathbf{E} = \frac{\omega^2}{c^2} \mathbf{E} . \tag{1.11}$$

The magnitude of the wave vector is therefore $|\mathbf{k}| = k = \omega/c$. The *index of refraction* n is related to the magnitude of the wave vector by $k = k_0 n$ where $k_0 = 2\pi/\lambda$ is the angular wavenumber. In a vacuum $n = 1$ and so $k = k_0$, in air, for all practical purposes, we may also set $n = 1$ otherwise $n > 1$.

Maxwell's equations in matter are mathematically identical to Eqs. (1.1–1.4) but with notational changes: μ_0, ϵ_0 are replaced by their values in matter, μ, ϵ while ρ, $\mathbf{J}$ are replaced by ρ_f, $\mathbf{J}_f$. In the absence of free charges $\rho_f = 0$. In the presence of free charges we can use Ohm's law which relates the current density to the electric field.

$$\mathbf{J}_f = \sigma \mathbf{E} , \tag{1.12}$$

where σ is the conductivity. Conservation of charge leads to the continuity equation

$$\nabla \cdot \mathbf{J}_f = -\frac{\partial \rho_f}{\partial t} . \tag{1.13}$$

Combining this with (1.12) and Gauss's Law, Eq. (1.1), leads to

$$\frac{\partial \rho_f}{\partial t} = -\frac{\sigma}{\epsilon} \rho_f \tag{1.14}$$

with solution

$$\rho_f(t) = \rho_f(0) e^{-(\sigma/\epsilon)t} . \tag{1.15}$$

This means that the free charge density ρ_f dissipates in a characteristic time $\tau = \epsilon/\sigma$ in a conductor [1]. Further, for sufficiently large t the exponential decay shows that the density $\rho_f(t)$ is negligible. For $t \gg \tau$ we may therefore ignore free charges and Maxwell's equations in matter are,

$$\nabla \cdot \mathbf{E} = 0 \tag{1.16}$$

$$\nabla \cdot \mathbf{B} = 0 \tag{1.17}$$

$$\nabla \times \mathbf{E} = -\frac{\partial \mathbf{B}}{\partial t} \tag{1.18}$$

$$\nabla \times \mathbf{B} = \mu\sigma\mathbf{E} + \mu\epsilon\frac{\partial \mathbf{E}}{\partial t}. \tag{1.19}$$

A wave equation may be obtained using the same method as in the earlier vacuum case but now we have to account for the current density. Following from (1.5) we now find

$$\nabla \times (\nabla \times \mathbf{E}) = -\frac{\partial}{\partial t}\nabla \times \mathbf{B} = -\frac{\partial}{\partial t}\left(\mu\sigma\mathbf{E} + \mu\epsilon\frac{\partial \mathbf{E}}{\partial t}\right). \tag{1.20}$$

Hence the new form of wave equation

$$\nabla^2\mathbf{E} = \mu\sigma\frac{\partial \mathbf{E}}{\partial t} + \mu\epsilon\frac{\partial^2 \mathbf{E}}{\partial t^2}. \tag{1.21}$$

Using the expression for $\mathbf{E}$, Eq. (1.9), we note $\partial\mathbf{E}/\partial t = -i\omega\mathbf{E}$ hence, due to the inclusion of the current density, the magnitude of the wave vector $k = |\mathbf{k}|$ is now complex,

$$k^2 = i\mu\sigma\omega + \mu\epsilon\omega^2. \tag{1.22}$$

We can split k into real and imaginary parts $k = k_r + ik_i$ where

$$k_r = \omega\sqrt{\frac{\epsilon\mu}{2}}\sqrt{s+1}, \quad k_i = \omega\sqrt{\frac{\epsilon\mu}{2}}\sqrt{s-1} \tag{1.23}$$

and

$$s = \sqrt{1 + \left(\frac{\sigma}{\epsilon\omega}\right)^2}. \tag{1.24}$$

We may write the vector $\mathbf{k} = k\hat{\mathbf{k}}$, where $\hat{\mathbf{k}}$ is the unit vector in the direction of propagation. Substituting for $\mathbf{k}$ into the expression for $\mathbf{E}$, Eq. (1.9), we find

$$\mathbf{E}(z,t) = \Re\left[\mathbf{E_m}e^{-k_i\widehat{\mathbf{k}}\cdot\mathbf{r}}e^{i(k_r\widehat{\mathbf{k}}\cdot\mathbf{r}-\omega t+\psi)}\right]$$

$$= \mathbf{E_m}e^{-k_i\widehat{\mathbf{k}}\cdot\mathbf{r}}\cos(k_r\widehat{\mathbf{k}}\cdot\mathbf{r} - \omega t + \psi). \tag{1.25}$$

This shows that as the electric field propagates it also decays exponentially with distance, where the characteristic distance or penetration depth is given by $1/k_i$.

A complex magnitude for the wave vector leads to a complex index of refraction

$$n = n_r + in_i \tag{1.26}$$

where

$$n_r = \frac{k_r}{k_0}, \qquad n_i = \frac{k_i}{k_0}. \tag{1.27}$$

1.3 Conclusions

When considering optics at the nanometer scale, the size of the optical wavelength and the beam diameter need to be taken into account. The overlap of incident and reflected light leads to optical interference on surfaces resulting in nonuniform optical fields near these surfaces. The interaction of light with matter can be described by light scattering where the medium is considered as an ensemble of scatterers.

The wave nature of the light may be derived from Maxwell's equations. The solutions of the wave equation lead to expressions for planar waves, where the electric and magnetic fields can be written as a function of the amplitude, polarization, the angular frequency ω and the wave vector $\mathbf{k}$. A useful feature of the wave equation is that it is linear. This means that if we have more than one solution, say $\mathbf{E}_1$, $\mathbf{E}_2$ then the sum $\mathbf{E} = \mathbf{E}_1 + \mathbf{E}_2$ is also a solution (this feature will be exploited in subsequent chapters).

In the following we will focus on the electric field of planar waves, using the solutions obtained in this chapter. The magnetic field is easily derived from the electric field and so, in general, we will not calculate this. The speed of propagation depends on the wave vector $\mathbf{k}$ which is proportional to the index of refraction. In matter the index of refraction n is complex; the real part n_r describes the speed of propagation of the wave and the complex part n_i describes the exponential decrease of the amplitude of the wave.

References

1. D.J. Griffiths, *Introduction to Electrodynamics* (Pearson, London, 2018). ISBN 978-93-325-5044-5
2. R.P. Feynman, R.B. Leighton, M. Sands, *The Feynman Lectures in Physics*, vol. 1 (Addison-Wesley Publishing Company, Menlo Park, 1963), pp. 31–33
3. V.F. Weisskopf, Sci. Am. **219**(3), 60 (1968)
4. W. Bacsa, *Interference Scanning Optical Probe Microscopy: Principles and Applications.* Advances in Imaging and Electron Physics, vol. 10 (Academic Press, Cambridge, 1999), pp. 1–19

Chapter 2
Optical Interference Near Surfaces: Interference Substrates

2.1 Overlapping Monochromatic Beams

Whenever two coherent light beams of the same wavelength overlap they interfere and give rise to optical standing waves, which cause interference fringes when recording images. If the light beam travels in the direction $\mathbf{k}$, the superposition of two beams with wave vectors $\mathbf{k}_1$ and $\mathbf{k}_2$ results in a wave that travels along $\mathbf{k}_1 + \mathbf{k}_2$ (see Fig. 2.1).

In the region of overlap, wave fronts from both the waves form a mesh that travels with time along the direction of $\mathbf{k}_1 + \mathbf{k}_2$. The points of crossing of the wave fronts of the two beams where they superimpose also move in the same direction. This means that the field maxima and minima are separated in space along the direction of propagation and form a standing wave which, when averaged over time, creates fringes aligned with $\mathbf{k}_1 + \mathbf{k}_2$. Figure 2.2 shows two overlapping monochromatic and coherent beams, illustrating the formation of standing waves. A similar overlap can be found for a beam reflected off a surface, as shown in Fig. 1.1. Here, the incident and reflected beams overlap to form an optical standing wave with fringes parallel to the substrate. The components of the wave vectors $\mathbf{k}_1$ and $\mathbf{k}_2$ oriented perpendicular to the surface oppose each other and cancel when taking the sum of $\mathbf{k}_1$ and $\mathbf{k}_2$ whereas the parallel components combine positively. In the following, we will refer to the standing wave generated by the incident and reflected wave as a surface standing wave.

The two beams overlap at a distance smaller than $d \cos(\theta)$ where d is the diameter of the beam and θ the angle of incidence. It is recalled here that when a light beam is incident on a surface, the beam that is reflected off this surface is generated by induced dipoles in the material. If we consider the component of the field of oscillating dipoles which propagates into the far field, we find that it is phase shifted by π. This has the consequence that for a dielectric material film, the interference at the substrate is destructive and the surface standing wave has a minimum there. Since a part of the incident wave is transmitted, the amplitude of the reflected wave is smaller than that of the incident wave and the field minimum of the surface

© The Author(s), under exclusive licence to Springer Nature Switzerland AG 2020
W. Bacsa et al., *Optics Near Surfaces and at the Nanometer Scale*,
SpringerBriefs in Physics, https://doi.org/10.1007/978-3-030-58983-7_2

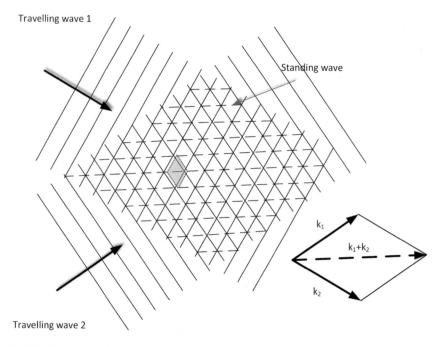

Fig. 2.1 Two overlapping monochromatic beams at a given instant. The overlapping fringe pattern is translated in the direction of the sum of the wave vectors of the two waves (dashed line). The time averaged field forms standing waves

standing wave is non-zero. This has the effect that observing optical signals from ultra-thin films ($<\lambda/10$) or monolayers of molecules or a single atomic layer such as graphene is particularly challenging because of the very small local field at the surface of the non-transparent substrate. However, if we add a transparent layer on top of the substrate, with thickness adjusted so that the interference maximum falls on its surface, it is possible to increase the optical field [1]. This occurs since the amplitude of the reflected beam from the transparent layer is small when compared to that reflected off the opaque substrate. In other words, by reducing the amplitude of the beam reflected off the transparent surface, we can increase the light intensity there. Transparent substrates are thus more favorable for optical observation than non-transparent ones and the local field may be maximized by combining an opaque substrate with a transparent layer whose thickness is adjusted to the wavelength of the incident beam. Taking advantage of this phenomenon, we can create interference substrates by combining a highly reflecting substrate with a thin transparent layer which has just the right thickness so that the interference maximum of the standing wave falls at the surface of the transparent layer. Using such intelligent interference substrates enables us to observe nanoparticles with a monochromatic beam rather than using an optical microscope in transmission mode. In the following section we look in more detail at the superposition of two waves and optical standing waves.

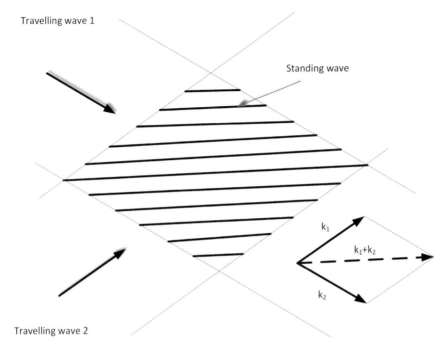

Travelling wave 1

Standing wave

k_1

k_1+k_2

k_2

Travelling wave 2

Fig. 2.2 When time averaging only the standing waves, interference fringes are recorded

2.2 Historical Note on the Observation of Optical Standing Waves

Optical standing waves were first observed by Wiener using thin films on a metal surface and a monochromatic light beam [2] to establish the wave nature of light. The formation of optical standing waves was proposed to explain the coloured images observed by E. Becquerel (1868). A silver plate coated with a thin layer of silver chloride coloured up under the action of light with the colour corresponding to that of the light used. The possible formation of standing waves was put forward by Zenker and Lord Rayleigh. Standing waves were used to create color images by Lippmann who earned the Nobel prize for this discovery in 1908. Ranson et al. [3] and Holm et al. [4] in their study on the photoluminescence (PL) of thin films reported that the PL signal oscillated as a function of the film thickness if the film was deposited on a reflecting substrate. This observation was explained to be due to multiple reflections and interference in the thin films. Similarly, when observing Raman spectral bands of nitrogen and oxygen layers on silver, oscillations in the Raman intensity were observed depending on the film thickness [5]. The oscillations were enhanced when the thickness of the film was an integer multiple of half the excitation wavelength when taking into account the index of refraction and the angle of incidence. Surface interference fringes as a function of angle of incidence were first observed with a scanning optical probe in collection mode by Umeda et al. [6]. Combining an interference substrate with a scanning optical probe in collection mode to obtain a high lateral resolution of 35 nm on Ag island films was later demonstrated [7]. Optical standing waves have been used to control a scanning optical probe near a liquid interface [8] and SiO_2 layers on Si have been used to enhance contrast of molecular layers in optical microscopy [9]. Interference substrates (SiO_2/Si) finally made it possible to observe graphene, a single layer of carbon, with an optical microscope [10].

2.3 Superposition of Two Plane Waves

In the following, we will write down analytical expressions for superimposed plane waves. A propagating plane wave is described by Eq. (1.8)

$$\mathbf{E} = \mathbf{E}_m \cos(\mathbf{k} \cdot \mathbf{r} - \omega t), \tag{2.1}$$

where, for simplicity, we have omitted the phase angle. The wave travels in the direction $\mathbf{k}$ which is normal to the amplitude vector $\mathbf{E}_m$, which gives the direction of the transverse field (polarisation).

In much of the following we will assume waves of the same angular frequency (for example if one wave is a reflection of the other or due to wave scattering) in which case the simplest way to understand the superposition of two waves is to start with the case where the waves $\mathbf{E}_1$, $\mathbf{E}_2$ have the same frequency ($\omega_1 = \omega_2 = \omega$) and equal amplitude vector ($\mathbf{E}_{m2} = \mathbf{E}_{m1} = \mathbf{E}_m$). Denoting $\phi = \mathbf{k} \cdot \mathbf{r} - \omega t$ we may express the sum of the waves as

$$\mathbf{E} = \mathbf{E}_1 + \mathbf{E}_2 = \mathbf{E}_m \left[\cos(\phi_1) + \cos(\phi_2) \right]. \tag{2.2}$$

Noting that $\cos(x) + \cos(y) = 2 \cos\left(\frac{x-y}{2}\right) \cos\left(\frac{x+y}{2}\right)$ this may be written

$$\mathbf{E} = 2\mathbf{E}_m \cos\left(\frac{\phi_1 - \phi_2}{2}\right) \cos\left(\frac{\phi_1 + \phi_2}{2}\right). \tag{2.3}$$

Substituting for ϕ_1, ϕ_2

$$\mathbf{E} = 2\mathbf{E}_m \cos\left[\frac{(\mathbf{k}_1 - \mathbf{k}_2) \cdot \mathbf{r}}{2}\right] \cos\left[\frac{(\mathbf{k}_1 + \mathbf{k}_2) \cdot \mathbf{r}}{2} - \omega t\right]. \tag{2.4}$$

Comparison with Eq. (2.1) shows that this is a wave propagating in the direction $\mathbf{k}_1 + \mathbf{k}_2$ whose amplitude is modulated perpendicular to it, that is, along $\mathbf{k}_1 - \mathbf{k}_2$. Since $\mathbf{E}_m$ is perpendicular to both $\mathbf{k}_1$, $\mathbf{k}_2$ it is also perpendicular to the new direction of propagation, $\mathbf{E}_m \cdot (\mathbf{k}_1 + \mathbf{k}_2) = \mathbf{E}_m \cdot \mathbf{k}_1 + \mathbf{E}_m \cdot \mathbf{k}_2 = 0$.

For the example of the superposition of an incident wave normal to a surface and its reflection, $\mathbf{k}_{in} = \mathbf{k}_1 =: \mathbf{k}$ and $\mathbf{k}_r = \mathbf{k}_2 = -\mathbf{k}_{in}$, then Eq. (2.4) reduces to

$$E = 2E_m \cos(\mathbf{k} \cdot \mathbf{r}) \cos(\omega t). \qquad (2.5)$$

This is the analytical description of a standing optical wave with amplitude $2E_m \cos(\mathbf{k.r})$ which oscillates at frequency ω.

The light intensity measured in an experiment (or registered by our eyes) is the time averaged power density or the energy flux per area and time. The energy flux is determined from the product of the electric and magnetic fields. To calculate this we begin with the directional energy flux which is termed the Poynting vector,

$$\mathbf{R} = \frac{1}{\mu_0} \mathbf{E} \times \mathbf{B}. \qquad (2.6)$$

For a wave of the form (2.1) it is easily shown that

$$\nabla \times \mathbf{E} = i\mathbf{k} \times \mathbf{E} \qquad (2.7)$$

and from Maxwell's equations (specifically (1.18))

$$\mathbf{k} \times \mathbf{E} = \omega \mathbf{B}. \qquad (2.8)$$

The Poynting vector may now be written

$$\mathbf{R} = \frac{1}{\mu_0 \omega} \mathbf{E} \times (\mathbf{k} \times \mathbf{E}). \qquad (2.9)$$

This may be simplified using the vector identity

$$\mathbf{u} \times (\mathbf{v} \times \mathbf{w}) = (\mathbf{u} \cdot \mathbf{w})\mathbf{v} - (\mathbf{u} \cdot \mathbf{v})\mathbf{w} \qquad (2.10)$$

so that

$$\mathbf{R} = \frac{1}{\mu_0 \omega} ((\mathbf{E} \cdot \mathbf{E})\mathbf{k} - (\mathbf{E} \cdot \mathbf{k})\mathbf{E}) = \frac{1}{\mu_0 \omega} (\mathbf{E} \cdot \mathbf{E})\,\mathbf{k}, \qquad (2.11)$$

since the directional amplitude is perpendicular to the wave vector. The magnitude of the directional energy flux may be considered as the product of the energy density U and the speed of the wave, $|\mathbf{R}| = Uc$, hence

$$U = \frac{1}{c\mu_0 \omega} (\mathbf{E} \cdot \mathbf{E}) \, |\mathbf{k}| = \epsilon_0 \mathbf{E} \cdot \mathbf{E} = \epsilon_0 |\mathbf{E}|^2, \qquad (2.12)$$

after noting that $|\mathbf{k}| = k = \omega/c$ and $c^2 = \epsilon_0 \mu_0$.

The time average of a quantity is defined by

$$\langle A \rangle = \frac{1}{T} \int_0^T A \, dt \,, \tag{2.13}$$

where T is the period of the wave. What is measured by instruments is the time average of the energy density which we may now write as

$$\langle U \rangle = \epsilon_0 \langle \mathbf{E} \cdot \mathbf{E} \rangle \tag{2.14}$$

with the period $T = 2\pi/\omega$.

For the special case of a wave travelling normal to a surface, defined by (2.5), we find

$$\langle \mathbf{E} \cdot \mathbf{E} \rangle = 4|\mathbf{E}_m|^2 \cos^2(\mathbf{k} \cdot \mathbf{r}) \frac{\omega}{2\pi} \int_0^{2\pi/\omega} \cos^2(\omega t) \, dt$$
$$= 2|\mathbf{E}_m|^2 \cos^2(\mathbf{k} \cdot \mathbf{r}) \,. \tag{2.15}$$

The maximum intensity of this field will be observed when $\cos^2(\mathbf{k} \cdot \mathbf{r}) = 1$, i.e. where $\mathbf{k} \cdot \mathbf{r} = \pi p$, where p is an integer. Without loss of generality we may orient the axes such that the wave propagates along z, so that $\mathbf{k} = k(0, 0, 1)$ then the planes of maximum intensity are defined by

$$\mathbf{k} \cdot \mathbf{r}_{max} = k \, z_{max} = \pi p \,. \tag{2.16}$$

Since the wavelength $\lambda = 2\pi/k$ in this case the fringe spacing, i.e. the distance between two consecutive planes (defined by p and $p + 1$) of maximum brightness, is just half of the wavelength

$$\delta_f = \frac{\pi}{k} = \frac{\lambda}{2n} \,, \tag{2.17}$$

where n is the index of refraction. If we consider waves travelling through air (or a vacuum) we may set $n = 1$. If the angle of incidence is not normal one needs to take the component of $\mathbf{k}_{in}$ perpendicular to the surface and it can be shown that

$$\delta_f = \frac{\lambda}{2n \cos(\theta_i)} \,. \tag{2.18}$$

where θ_i is the angle of incidence (the angle between $\mathbf{k}_{in}$ and the normal to the surface).

In practice it is more likely that the amplitude vector of the waves are not equal, then the sum

$$\mathbf{E} = \mathbf{E}_{m1} \cos(\phi_1) + \mathbf{E}_{m2} \cos(\phi_2) \,. \tag{2.19}$$

For the special case of a normal incident wave reflected off a surface $\mathbf{k}_1 = -\mathbf{k}_2 = \mathbf{k}$ we may exploit the results of the previous case by first splitting $\mathbf{E}_{m2}$ into two components and introducing a new vector $\mathbf{E}_{m3} := \mathbf{E}_{m1} - \mathbf{E}_{m2}$. This allows us to write

$$\mathbf{E} = \mathbf{E}_{m1}(\cos(\phi_1) + \cos(\phi_2)) - \mathbf{E}_{m3}\cos(\phi_2) \tag{2.20}$$

The $\mathbf{E}_{m1}$ term was dealt with in the analysis following Eq. (2.2) and so

$$\mathbf{E} = 2\mathbf{E}_{m1}\cos\left[\frac{\phi_1 - \phi_2}{2}\right]\cos\left[\frac{\phi_1 + \phi_2}{2}\right] - \mathbf{E}_{m3}\cos(\phi_2)$$
$$= 2\mathbf{E}_{m1}\cos\left[\frac{(\mathbf{k}_1 - \mathbf{k}_2) \cdot \mathbf{r}}{2}\right]\cos\left[\frac{(\mathbf{k}_1 + \mathbf{k}_2) \cdot \mathbf{r}}{2} - \omega t\right]$$
$$- (\mathbf{E}_{m1} - \mathbf{E}_{m2})\cos[\mathbf{k}_2 \cdot \mathbf{r} - \omega t] . \tag{2.21}$$

Using $\mathbf{k}_1 = -\mathbf{k}_2 = \mathbf{k}$, the following expression is obtained:

$$\mathbf{E} = 2\mathbf{E}_{m1}\cos(\mathbf{k} \cdot \mathbf{r})\cos(\omega t) - (\mathbf{E}_{m1} - \mathbf{E}_{m2})\cos[\mathbf{k} \cdot \mathbf{r} + \omega t] \tag{2.22}$$
$$= \mathbf{E}_{st} + \mathbf{E}_{pr} . \tag{2.23}$$

Previously, when the same amplitude vector waves were superimposed we obtained a standing wave. Now the amplitude vectors are different we obtain a combination of a standing wave and a propagating wave. Note that since the standing wave is unchanged from before, the fringe spacing δ_f of the standing wave is not influenced by the fact that the two amplitudes are different.

With arbitrary $\mathbf{k}$ and different amplitudes it is simpler to understand the superposition of two waves when using complex notation. In complex form we may write

$$E_1 = \mathbf{E}_{m1}\cos(\mathbf{k}_1.\mathbf{r} - \omega t) = \Re\left(\mathbf{E}_{m1}e^{i\phi_1}\right) \tag{2.24}$$
$$E_2 = \mathbf{E}_{m2}\cos(\mathbf{k}_2.\mathbf{r} - \omega t) = \Re\left(\mathbf{E}_{m2}e^{i\phi_2}\right) . \tag{2.25}$$

It is important to state that since the wave is a real quantity at the end of the calculations we only require the real part, hence the $\Re$ notation. To make the mathematics simpler it is standard at this stage to focus on a single component of the $\mathbf{E}_i$ vectors, the analysis is exactly the same for all other components. Consequently we now write

$$E_1 = E_{m1}\cos(\mathbf{k}_1.\mathbf{r} - \omega t) = \Re\left(E_{m1}e^{i\phi_1}\right) \tag{2.26}$$
$$E_2 = E_{m2}\cos(\mathbf{k}_2.\mathbf{r} - \omega t) = \Re\left(E_{m2}e^{i\phi_2}\right) . \tag{2.27}$$

Since the wave equation is linear we may add the two to form a third wave which, also being a wave, must be expressible in the same form

$$E = \Re\left(E_m e^{i\Phi}\right) \tag{2.28}$$

Fig. 2.3 Illustration of the addition of two complex numbers

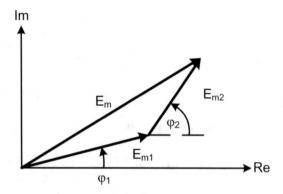

for some, as yet unknown, amplitude E_m and angle Φ. From now on we will omit the real notation, with the understanding that at the end of the calculation it is only the real part of the wave that is needed. To determine E_m, Φ we refer to Fig. 2.3. Resolving each wave into its real and imaginary components we see that the sum of E_1 and E_2 is

$$
\begin{aligned}
E &= (E_{m1} \cos \phi_1 + E_{m2} \cos \phi_2) + i(E_{m1} \sin \phi_1 + E_{m2} \sin \phi_2) \\
  &= E_m (\cos \Phi + i \sin \Phi) \; .
\end{aligned}
\tag{2.29}
$$

From the figure the amplitude E_m is found immediately via the cosine rule

$$
\begin{aligned}
E_m^2 &= E_{m1}^2 + E_{m2}^2 - 2E_{m1} E_{m2} \cos(\pi - \delta) \; , \\
      &= E_{m1}^2 + E_{m2}^2 + 2E_{m1} E_{m2} \cos \delta \; ,
\end{aligned}
\tag{2.30}
$$

where the angle between the vectors is $\pi - \delta$ and $\delta = \phi_2 - \phi_1 = (\mathbf{k}_2 - \mathbf{k}_1).\mathbf{r}$. The angle of E may be defined through the definition (2.28)

$$
\cos \Phi = \frac{\Re(E)}{E_m} = \frac{E_{m1} \cos \phi_1 + E_{m2} \cos \phi_2}{E_m} \; .
\tag{2.31}
$$

Using the equality

$$
\cos(a - b) = \cos a \cos b + \sin a \sin b
\tag{2.32}
$$

and the definitions of ϕ_1, ϕ_2 we may expand Eq. (2.31) to isolate the time-dependent part

$$
\begin{aligned}
E_m \cos \Phi &= E_{m1} \cos \phi_1 + E_{m2} \cos \phi_2 \\
              &= \cos \omega t \, [E_{m1} \cos (\mathbf{k}_1 \cdot \mathbf{r}) + E_{m2} \cos (\mathbf{k}_2 \cdot \mathbf{r})] \\
              &\quad + \sin \omega t \, [E_{m1} \sin (\mathbf{k}_1 \cdot \mathbf{r}) + E_{m2} \sin (\mathbf{k}_2 \cdot \mathbf{r})] \; .
\end{aligned}
\tag{2.33}
$$

We define a time-independent phase angle, Φ', from this equation such that

$$E_m \cos \Phi' = E_{m1} \cos (\mathbf{k}_1 \cdot \mathbf{r}) + E_{m2} \cos (\mathbf{k}_2 \cdot \mathbf{r}) \tag{2.34}$$

$$E_m \sin \Phi' = E_{m1} \sin (\mathbf{k}_1 \cdot \mathbf{r}) + E_{m2} \sin (\mathbf{k}_2 \cdot \mathbf{r}) \ . \tag{2.35}$$

Then we may write Eq. (2.33) as

$$E_m \cos \Phi = E_m \left(\cos \omega t \cos \Phi' + \sin \omega t \sin \Phi' \right)$$
$$= E_m \cos(\Phi' - \omega t) \ . \tag{2.36}$$

So, the phase of the resultant wave may be written as $\Phi = \Phi' - \omega t$. Since our goal from the start was to determine $E = \Re(E_m e^{i\Phi}) = E_m \cos \Phi$ our task is complete: E_m is defined by Eq. (2.30), Φ' (and hence Φ) from either of Eqs. (2.34, 2.35). Further, at no stage did we specify which component of $\mathbf{E}$ we were dealing with, so the calculation holds for any component of the vector.

Our true goal is to determine the time average of the energy density so as to interpret interference patterns. For a wave composed of the two individual waves (2.24, 2.25) this is

$$\langle U \rangle = \epsilon_0 \langle (\mathbf{E}_1 + \mathbf{E}_2) \cdot (\mathbf{E}_1 + \mathbf{E}_2) \rangle = \epsilon_0 \langle \mathbf{E}_1 \cdot \mathbf{E}_1 \rangle + \epsilon_0 \langle \mathbf{E}_2 \cdot \mathbf{E}_2 \rangle + 2\epsilon_0 \langle \mathbf{E}_1 \cdot \mathbf{E}_2 \rangle$$
$$= \epsilon_0 |\mathbf{E}_{m1}|^2 \langle \cos^2 \phi_1 \rangle + \epsilon_0 |\mathbf{E}_{m2}|^2 \langle \cos^2 \phi_2 \rangle + 2\epsilon_0 \langle \mathbf{E}_1 \cdot \mathbf{E}_2 \rangle$$
$$= \langle U_1 \rangle + \langle U_2 \rangle + 2\epsilon_0 \langle \mathbf{E}_1 \cdot \mathbf{E}_2 \rangle \ . \tag{2.37}$$

The first two terms on the right hand side are simply the time averaged energy densities of the two individual waves.

$$\langle U_1 \rangle = \epsilon_0 |\mathbf{E}_{m1}|^2 \langle \cos^2 \phi_1 \rangle = \epsilon_0 |\mathbf{E}_{m1}|^2 \frac{\omega}{2\pi} \int_0^{2\pi/\omega} \cos^2 (\mathbf{k}_1 \cdot \mathbf{r} - \omega t) \, dt$$
$$= \frac{\epsilon_0 |\mathbf{E}_{m1}|^2}{2} \tag{2.38}$$

and similarly for $\langle U_2 \rangle$. The final term represents the interference between the two: this is key to understanding the interference patterns observed in experiments

$$2\epsilon_0 \langle \mathbf{E}_1 \cdot \mathbf{E}_2 \rangle = 2\epsilon_0 |\mathbf{E}_{m1}||\mathbf{E}_{m2}| \frac{\omega}{2\pi} \int_0^{2\pi/\omega} \cos(\mathbf{k}_1 \cdot \mathbf{r} - \omega t) \, \cos(\mathbf{k}_2 \cdot \mathbf{r} - \omega t) \, dt$$
$$= \epsilon_0 |\mathbf{E}_{m1}||\mathbf{E}_{m2}| \cos((\mathbf{k}_1 - \mathbf{k}_2) \cdot \mathbf{r}) \ . \tag{2.39}$$

Hence the energy density resulting from two interacting, propagating waves may be written

$$\langle U \rangle = \langle U_1 \rangle + \langle U_2 \rangle + 2\epsilon_0 \langle \mathbf{E}_1 \cdot \mathbf{E}_2 \rangle$$

$$= \frac{\epsilon_0}{2} \left\{ |\mathbf{E}_{m1}|^2 + |\mathbf{E}_{m2}|^2 + 2|\mathbf{E}_{m1}||\mathbf{E}_{m2}| \cos((\mathbf{k}_1 - \mathbf{k}_2) \cdot \mathbf{r}) \right\} . \tag{2.40}$$

In the special case where $|\mathbf{E}_{m1}| = |\mathbf{E}_{m2}|$ and $\mathbf{k}_1 = -\mathbf{k}_2$ we retrieve Eq. (2.15) (after applying $\cos(2a) + 1 = 2\cos^2 a$).

The maximum positive interference (and hence the brightest parts of an interference pattern) occurs when $\langle U \rangle$ is a maximum, which corresponds to $\cos((\mathbf{k}_1 - \mathbf{k}_2) \cdot \mathbf{r}) = 1$, that is when $(\mathbf{k}_1 - \mathbf{k}_2) \cdot \mathbf{r} = \pm 2p\pi$, where p is an integer. The darkest parts of an interference pattern occur when $(\mathbf{k}_1 - \mathbf{k}_2) \cdot \mathbf{r} = \pm(2p - 1)\pi$. If we allow for a phase shift, as discussed in the subsequent chapter then $\mathbf{k}_i \cdot \mathbf{r} - \omega t$ is replaced by $\mathbf{k}_i \cdot \mathbf{r} - \omega t + \psi_i$, where ψ_i is the phase. The analysis follows in exactly the same manner as above, and so we simply need to replace ϕ_i with $\phi_i + \psi_i$. We may define $\psi_1 = 0$ (so then we measure the shift from the first wave) and then the phase shift $\psi = \psi_2 - \psi_1 = \psi_2$. Including the phase the maximum positive interference occurs when $(\mathbf{k}_1 - \mathbf{k}_2) \cdot \mathbf{r} + \psi = \pm 2p\pi$, the darkest parts occur when $(\mathbf{k}_1 - \mathbf{k}_2) \cdot \mathbf{r} + \psi = \pm(2p - 1)\pi$ [11, 12].

If we wish to analyse an interference pattern (such as those shown in subsequent chapters) it is sufficient to simply explore the variation of $\mathbf{E}_1 \cdot \mathbf{E}_2$ and, more specifically, $(\mathbf{k}_1 - \mathbf{k}_2) \cdot \mathbf{r}$.

When the two amplitudes are not the same the amplitude of the interference term is reduced and one can define a fringe visibility,

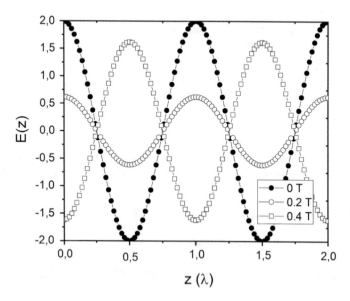

Fig. 2.4 The electric field of two counter propagating waves of equal amplitudes at three instants in time at $t = 0$, $t = 0.2T$ and $t = 0.4T$, where T is the period of the wave. Maxima and minima occur at the same z value

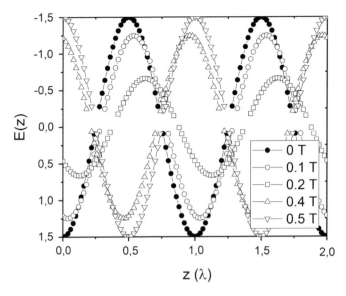

Fig. 2.5 The electric field of two counter propagating waves of non-equal amplitudes $E_{m1} = 1$, $E_{m2} = 0.5$ at five instants in time. Maxima and minima do not occur at the same z value

$$v := \frac{\langle U \rangle_{max} - \langle U \rangle_{min}}{\langle U \rangle_{max} + \langle U \rangle_{min}} = \frac{2|\mathbf{E}_{m1}||\mathbf{E}_{m2}|}{|\mathbf{E}_{m1}|^2 + |\mathbf{E}_{m2}|^2} . \tag{2.41}$$

Figure 2.4 shows the electric field as a function of z at three instants in time. Maxima and minima are located every half a wavelength in the case where the two waves propagate in opposite directions. Figure 2.5 shows the electric field as a function of z at five instants in time. Maxima and minima are not located at the same place or phase shifted but the amplitude is modulated in the same way as in the case of equal amplitudes.

2.4 Interference Substrate and the Electric Field at Its Surface

Here we describe how using an appropriately tailored interference substrate allows the electric field to be maximised at its surface and how this effect can be exploited in spectroscopy and microscopy. Since the incident and reflected waves are out of phase their interference at the surface is destructive and so the optical standing wave has a minimum there. This is a particularly unfavorable condition for the study of thin surface layers or the substrate itself with optical waves. The amplitude of the standing wave is highest when the reflected wave has a large amplitude or the visibility, defined by Eq. (2.41), $v = 1$. We have seen earlier that the deposition of an ultra-thin film on

a reflecting surface has the effect that the interference maximum falls on the surface
which serves to amplify the optical signal from molecular layers on its surface.
The first interference maximum occurs at half the fringe spacing $\lambda/(4n \cos \theta_i)$, see
Eq. (2.18), where n is the index of refraction of the transparent layer. The thickness
of the transparent layer may then be written

$$d = \left(p + \frac{1}{2}\right) \frac{\lambda}{2n \cos(\theta_i)} \tag{2.42}$$

where p is an integer. This demonstrates that we may use layers of different thickness
depending upon the choice of p. The index of refraction reduces the thickness of this
layer but does not influence the optical standing wave since only a small fraction is
reflected off at its surface. What is important here is that the field enhancement is
obtained without the use of electronic resonance; moreover the interference substrate
is simple to fabricate. Thus using an interference substrate with a SiO_2 layer it is
possible to detect optical spectra from molecular mono layers [1].

In the following, we determine the electric field at the interference substrate while
taking into account transmission and reflection, as well the angle of incidence at the
two interfaces of the interference substrate.

In order to calculate the amplitude at the surface of a transparent layer on an
opaque substrate we need to know the amplitudes of the incident and the reflected
beams at the air interface [12]. The incident wave is given by

$$\mathbf{E}(z, t) = \mathbf{E}_m \exp(i(kz - \omega t)). \tag{2.43}$$

For a plane wave travelling along $+z, k = k_0$ and when travelling along $-z, k = -k_0$.
The incident and reflected waves propagate in opposite directions, which we define
as $\mathbf{E}^+$ and $\mathbf{E}^-$

$$\mathbf{E}^+ = \mathbf{E}_m \exp(i(k_0 z - \omega t)), \tag{2.44}$$
$$\mathbf{E}^- = \mathbf{E}_m \exp(-i(k_0 z + \omega t)). \tag{2.45}$$

The plane wave in the $+z$ direction can be rewritten

$$\mathbf{E}(z, t) = \mathbf{E}_m \exp(-i\omega t) \exp(ik_0 z). \tag{2.46}$$

Defining

$$\mathbf{E_m}(t) = \mathbf{E}(0, t) = \mathbf{E}_m \exp(-i\omega t) \tag{2.47}$$

we obtain

$$\mathbf{E}(z, t) = \mathbf{E}_m(t) \exp(ik_0 z) \tag{2.48}$$

and so

$$\mathbf{E}^+(z, t) = \mathbf{E}_m(t) \exp(ik_0 z), \qquad \mathbf{E}^-(z, t) = \mathbf{E}_m(t) \exp(-ik_0 z). \qquad (2.49)$$

When the wave travels a distance d

$$\mathbf{E}^+(d, t) = \mathbf{E}_m(t) \exp(ik_0 d) \qquad \mathbf{E}^-(d, t) = \mathbf{E}_m(t) \exp(-ik_0 d). \qquad (2.50)$$

The two amplitudes at $z = 0$ are related to the amplitudes at $z = d$ by

$$\mathbf{E}^+(0, t) = \exp(-ik_0 d)\mathbf{E}^+(d, t) \qquad \mathbf{E}^-(0, t) = \exp(ik_0 d)\mathbf{E}^-(d, t). \qquad (2.51)$$

This may be rewritten

$$\begin{bmatrix} \mathbf{E}^+(0, t) \\ \mathbf{E}^-(0, t) \end{bmatrix} = L(d) \begin{bmatrix} \mathbf{E}^+(d, t) \\ \mathbf{E}^-(d, t) \end{bmatrix} \qquad (2.52)$$

where $L(d)$ is

$$L(d) = \begin{bmatrix} \exp(-ik_0 d) & 0 \\ 0 & \exp(ik_0 d) \end{bmatrix}. \qquad (2.53)$$

This matrix L is called the **propagation matrix**. It relates the amplitudes of the two propagating waves at two locations in one medium.

When a light wave crosses the interface between two materials labelled 1 and 2 (as shown in Fig. 2.6), the following condition for the electric field needs to be satisfied: the component of the electric field parallel to the interface has to be the same on both sides of the interface. This is a direct consequence of Ampere's law that describes the electric field generated by an electromagnetic wave as it crosses an interface. The amplitude of the wave in material 2 travelling in the $+z$ direction is the amplitude in material 1 travelling along $+z$ times the transmission coefficient plus the amplitude of the reflected wave travelling along $-z$ times the reflection coefficient, hence

Fig. 2.6 Amplitudes of waves across an interface travelling along +z and -z. The arrows indicate the direction of propagation. Note that the electric field vectors point along y

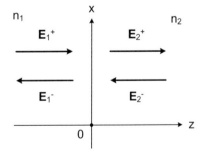

$$\mathbf{E}_2^+ = \tau_{12}\mathbf{E}_1^+ + \rho_{21}\mathbf{E}_2^- \tag{2.54}$$

where τ_{12} and ρ_{21} are the respective transmission and reflection coefficients. Figure 2.6 shows the four amplitudes in the two media. Similarly the amplitude of the wave in material 1 travelling along $-z$ is

$$\mathbf{E}_1^- = \tau_{21}\mathbf{E}_2^- + \rho_{12}\mathbf{E}_1^+ . \tag{2.55}$$

From (2.54) we may express $\mathbf{E}_1^+$ in terms of the field in material 2,

$$\mathbf{E}_1^+ = \frac{\mathbf{E}_2^+ - \rho_{21}\mathbf{E}_2^-}{\tau_{12}} = \frac{\mathbf{E}_2^+ + \rho_{12}\mathbf{E}_2^-}{\tau_{12}} , \tag{2.56}$$

after noting that the reflection coefficients $\rho_{21} = -\rho_{12}$. This can be used in (2.55) to determine $\mathbf{E}_1^-$ in terms of the field in material 2

$$\mathbf{E}_1^- = \frac{\tau_{21}\tau_{12} + \rho_{21}\rho_{12}}{\tau_{12}}\mathbf{E}_2^- + \frac{\rho_{12}}{\tau_{12}}\mathbf{E}_2^+ \tag{2.57}$$

$$= \frac{1}{\tau_{12}}\mathbf{E}_2^- + \frac{\rho_{12}}{\tau_{12}}\mathbf{E}_2^+ \tag{2.58}$$

after noting that $\tau_{21}\tau_{12} + \rho_{21}\rho_{12} = 1$ (since $T + R = 1$ and $T := \tau_{21}\tau_{12}$, $R := \rho_{21}\rho_{12}$). We have therefore obtained the amplitudes of the two waves in medium 1 in terms of the two amplitudes in medium 2

$$\begin{bmatrix} \mathbf{E}_1^+ \\ \mathbf{E}_1^- \end{bmatrix} = H_{12}\begin{bmatrix} \mathbf{E}_2^+ \\ \mathbf{E}_2^- \end{bmatrix} , \tag{2.59}$$

where the **interface matrix** H_{12} is

$$H_{12} = 1/\tau_{12}\begin{bmatrix} 1 & \rho_{12} \\ \rho_{12} & 1 \end{bmatrix} . \tag{2.60}$$

With the two matrices L and H the amplitudes of the waves propagating in the $+z$ and $-z$ directions across an interface and in the material can be calculated. The advantage of introducing these matrices is that it is then a simple matter to relate amplitudes across several layers. For example, when considering only a substrate, there is only one interface, that between air and the substrate and so

$$\begin{bmatrix} \mathbf{E}_1^+ \\ \mathbf{E}_1^- \end{bmatrix} = L_{12}H_{12}\begin{bmatrix} \mathbf{E}_2^+ \\ \mathbf{E}_2^- \end{bmatrix} . \tag{2.61}$$

When considering a single layer on a substrate there are two materials, the substrate and the layer in addition to air (index 0):

$$\begin{bmatrix} E_0^+ \\ E_0^- \end{bmatrix} = L_{01} H_{01} L_{12} H_{12} \begin{bmatrix} E_2^+ \\ E_2^- \end{bmatrix}.$$

(2.62)

For each layer two additional matrices L and H need to be added or in general, the total matrix, S which is given by:

$$S_{0N} := L_{01} H_{01} .. L_{N-1N} H_{N-1N}$$

(2.63)

and the amplitudes may be expressed using S:

$$\begin{bmatrix} E_0^+ \\ E_0^- \end{bmatrix} = S_{0N} \begin{bmatrix} E_N^+ \\ E_N^- \end{bmatrix}.$$

(2.64)

The matrix S thus includes the effect of all the layers and also takes into account multiple reflections. To calculate S we need to know ρ, τ, d, n for each layer and for the substrate.

To determine the Fresnel coefficient for ρ and τ we first define a and b:

$$a := \frac{\cos(\alpha_2)}{\cos(\alpha_1)} \qquad b := \frac{n_2}{n_1}.$$

(2.65)

When the incident light is polarized perpendicular to the plane of incidence (s polarization):

$$\tau_s := \frac{2}{1 + ab} \qquad \rho_s := \frac{1 - ab}{1 - ab}.$$

(2.66)

When polarized parallel to the plane of incidence (p polarization):

$$\tau_p := \frac{2}{a + b} \qquad \rho_p := \frac{1 - a/b}{1 + a/b}.$$

(2.67)

For **absorbing materials** the index of refraction is complex $n_2 = n_{r2} + i n_{i2}$ and consequently so is the wave vector. If we denote $\mathbf{k}_2 = \mathbf{k}_{r2} + i \mathbf{k}_{i2}$ then tangential continuity at the boundary between the atmosphere and absorbing material requires

$$\mathbf{k}_1 \cdot \hat{\mathbf{t}} = \mathbf{k}_2 \cdot \hat{\mathbf{t}}$$

(2.68)

where $\mathbf{k}_1$ represents the incoming wave and, if the interface is oriented along $\mathbf{e}_z$, then $\hat{\mathbf{t}} = (0, 1, 0)$ is the unit tangent at the interface. The left hand side is real while the right hand side contains a complex vector. Comparing real and imaginary parts we obtain

$$\mathbf{k}_1 \cdot \hat{\mathbf{t}} = \mathbf{k}_{r2} \cdot \hat{\mathbf{t}} \quad \Rightarrow \quad k_1 \sin(\phi_1) = k_2 \sin(\phi_2)$$

(2.69)

$$\mathbf{k}_{i2} \cdot \hat{\mathbf{t}} = 0.$$

(2.70)

The first equation is obviously just a statement of Snell's law, which we may also write in terms of the indices of refraction, $n_1 \sin(\phi_1) = n_2 \sin(\phi_2)$. The second equation demonstrates that, at the boundary, $\mathbf{k}_{i2}$ is normal to the boundary. So $\mathbf{k}_{r2}$ and $\mathbf{k}_{i2}$ have different directions [13].

Now, we move away from the boundary and examine the electric field in the material

$$E_2 = E_{m2} \exp(i(\mathbf{k}_2 \cdot \mathbf{r} - \omega t)). \tag{2.71}$$

Substituting

$$\mathbf{k}_2 \cdot \mathbf{r} = n_2 k_0 (0, \sin(\phi_2), \cos(\phi_2)) \cdot (x, y, z) \tag{2.72}$$

into the field equation

$$E_2 = E_m \exp(i(n_2 k_0 (\sin(\phi_2)y + \cos(\phi_2)z) - \omega t)). \tag{2.73}$$

We may replace the y coefficient via Snell's law and the phase factor is then

$$\exp(i(k_0(n_1 \sin(\phi_1)y + n_2 \cos(\phi_2)z) - \omega t)). \tag{2.74}$$

Substituting for n_2 leads to

$$\exp(i(k_0(n_1 \sin(\phi_1)y + n_{r2} \cos(\phi_2)z) - \omega t)) \exp(-n_{i2} k_0 \cos(\phi_2)z). \tag{2.75}$$

The above relation indicates that the wave in medium 2 falls off exponentially along z according to

$$\exp(-n_{i2} k_0 \cos(\phi_2)z) \tag{2.76}$$

where $1/(n_{i2} k_0 \cos(\phi_2))$ is the penetration depth. The part that propagates is along

$$\mathbf{k}_2 = k_0 \begin{pmatrix} 0 \\ n_1 \sin(\phi_1) \\ n_{r2} \cos(\phi_2) \end{pmatrix}. \tag{2.77}$$

So, for absorbing materials the angle of refraction is given by the real part of the complex index of refraction and the refracted wave falls off exponentially in a direction perpendicular to the interface.[1] In Sect. 5.3.1 we discuss the index of refraction and decay in the context of the retarded time, which takes into account the finite speed of light.

---

[1] This is not strictly correct. An effective index of refraction can be introduced which depends on the angle of incidence. The final result is, however, the same within a few percent for most values for n_{r2} and n_{i2} [12].

The amplitudes for waves in both directions and on both sides of the interface can be calculated in terms of the matrix S which is given by the reflection and transmission coefficients ρ and τ, while ρ and τ depend on the index of refraction n_1 and n_{r2}.

Let us consider the Si/air interface. As an example the electric field in the proximity of the surface of crystalline silicon ($n_2 = 4.42 + 0.9i$) will now be calculated. The interface is oriented along $+z$. The material fills the half space $z < 0$ (index 2), and air $z > 0$ (index 1). We first assume the incident field propagates towards the interface ($+$ direction) in the material at the distance of d_2 from the interface and calculate the field at the interface using the propagation matrix L where d is replaced by z and k_0 is replaced by k_2

$$\begin{bmatrix} E_2^+ \\ E_2^- \end{bmatrix} = L(k_2, z) \begin{bmatrix} 1 \\ 0 \end{bmatrix}. \tag{2.78}$$

Which yields

$$E_2^+(z) = \exp(-ik_{2z}z)E_2^+(0). \tag{2.79}$$

The amplitude on the air side can then be deduced by multiplying the interface matrix and the propagation matrix with the field obtained in the substrate at the interface.

$$\begin{bmatrix} E_1^+ \\ E_1^- \end{bmatrix} = L(k_1, z) H(\rho, \tau) \begin{bmatrix} E_2^+(d_2) \\ 0 \end{bmatrix}. \tag{2.80}$$

Both depend on α_2 which can be found using n_{2r} and Snell's law of refraction. The amplitude needs to be finally normalized with respect to the amplitude of the incident wave $E_1^-(d_2)$.

Figure 2.7 shows the square of the sum of the amplitudes of the incident and reflected waves

$$E_1^2(z) = (E_1^+(z) + E_1^-(z))^2. \tag{2.81}$$

and

$$E_2^2(z) = (E_2^+(z))^2 \tag{2.82}$$

The standing waves are formed at the side of the incident beam $+z$ with a fringe visibility smaller than one. In the case of normal incidence, the same result is obtained by considering two beam interference in Sect. 2.3 using $E_{m1} = 1$ and $E_{m2} = \rho_{21} E_{m1}$.

The standing wave thus depends on the angle of incidence. Fringe spacing increases with increasing angle and the intensity maximum is higher due to a higher amplitude of reflection. Note that the amplitude at the interface is only a small frac-

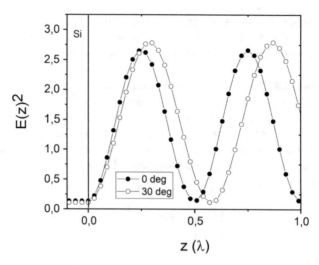

Fig. 2.7 Square of the electric field across a Si/air interface as a function of angle of incidence

tion of the amplitude of the interference maximum. The fringe contrast is 0.92 and the first interference fringe is at

$$\frac{\lambda}{4n_{SiO_2}\cos(\phi)}. \tag{2.83}$$

If we add a SiO_2 layer on the Si substrate and consider reflection at this additional interface and the propagation of incident and reflected beams using the S matrix: $S(i, j) = H(i, j)L(d, k)$ we can write the field in the SiO_2 layer as

$$\mathbf{E_1}(z) = S(1, 2)\,\mathbf{E_2}(d_2). \tag{2.84}$$

The field in air is

$$\mathbf{E_0}(z) = S(0, 1)\,\mathbf{E_1}(d_1) \tag{2.85}$$

where d_1 is the thickness of the SiO_2 layer and d_2 is the position of the Si/SiO_2 interface.

The optimal thickness of the SiO_2 layer is 85.6 nm for $\lambda = 500$ nm. The squared field on the SiO_2 surface is thus 16.7 times larger than on Si and 1.8 times larger than the maximum amplitude in air. The higher field at the surface of the interference substrate makes it particularly useful to optically observe ultra thin films such as molecular layers or nanoparticles. Figure 2.8 shows the square of the electric field in the substrate, SiO_2 layer and air at normal incidence and at an angle of 30 deg. This shows that the exact position of the interference maximum depends again on the angle of incidence.

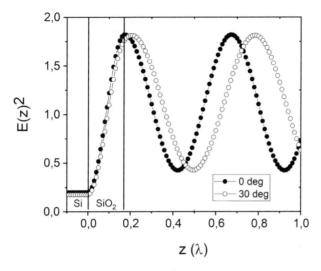

Fig. 2.8 Square of the electric field across Si/SiO$_2$/air interference substrate, at normal incidence and at 30°

2.5 Applications of Interference Substrates

In the previous section, we described how the electric field across a Si/SiO$_2$/air interference substrate can be enhanced by adjusting the SiO$_2$ layer thickness for a given wavelength of light and how this effect can be exploited to optically image nanoparticles on an appropriately tailored substrate.

The first detailed study of the interference effect and its application to spectroscopy was reported by Bacsa and Lannin [1] who exploited this effect to enhance Raman signals from nanomaterials by developing a new technique which they named BIERS (bilayer interference enhanced Raman spectroscopy). The Raman process, due to its second order nature, is associated with a low photon scattering efficiency which limits its sensitivity particularly in nanometre scale systems, such as ultra-thin films and clusters. In BIERS, the interference of incident and reflected beams from a dielectric/metal bilayer is used to enhance the exciting electric field. In the case of a SiO$_2$/Al bilayer, an interference enhancement by a factor of 27 could be obtained under an exciting wavelength of 514.5 nm and for SiO$_2$ thickness of 62 nm. This enhancement was experimentally demonstrated by comparing the in situ multichannel Raman signal of one monolayer of C_{60} on a bare Al substrate to that deposited on a SiO$_2$/Al bilayer. Since a thin SiO$_2$ layer has only a weak Raman signal and the neighboring Al layer has no first-order Raman active modes, the combination of the two layers was found to be a very suitable substrate to study ultrathin layers by Raman spectroscopy. The BIERS technique is widely used today as a versatile tool to enhance the sensitivity of Raman spectroscopy of nanoscale systems and bilayer substrates are widely available or can be custom made on demand. In a modified version of the BIERS substrate, plasmonic substrates containing subsurface gratings made of Ag nanoparticle assemblies were reported by Carles et al. [14]. By simultaneously exploiting optical interference and plasmonic resonance in these substrates, both elastic and inelastic scattering were enhanced. Sukmanowski et al. [15] used Raman spectroscopy to probe local electric fields in nanometre-thick multilayers of a-Si, Al, Ag and Al_2O_3. They found that it is possible to enhance the electrical field in a thin a-Si layer by using an amplifying layer that was tailored to exploit optical interference and the SERS effect to maximize the absorption and concentrate the incident light field on the active part of an a-Si thin film solar cell. Their results showed that in the presence of the amplifying layer, absorption can be considerably increased in thin Si layers (<50 nm thickness).

Two dimensional (2D) materials are atomically thin solids which have a strong in-plane covalent bonding and interact with other layers through van der Waals forces. 2D materials are hot topics of research in view of their interesting properties arising from their unique structure and have potential applications in various fields [16]. Today, the availability of interference substrates has made optical imaging of 2D materials a reality. Optical imaging is the first step in the characterisation of large-area atomic layers to rapidly gain global information on the general morphological characteristics and to detect whether the sample is a monolayer or whether multilayers are present. In terms of complexity and cost, optical imaging has obvious advantages over SEM (which is not suitable for large area samples) or AFM (due to its complexity of operation). However, the optical absorbance of an atomic layer such as mechanically exfoliated graphene is very low in the visible range (of the order of 2.3% per single layer). In addition, monolayers are deposited on a substrate which also absorbs light, but using interference substrates such SiO_2 or Si_3N_2 coated silicon wafers, high-contrast optical imaging of graphene under reflective illumination has been demonstrated [17]. The thickness of the dielectric layer is optimized depending on the illuminating wavelength. It is on such a substrate that graphene was first characterised by Novoselov et al. [10] in 2004 using an optical microscope under reflective illumination. The effective refractive index and optical absorption coefficient of graphene oxide, thermally reduced graphene oxide, and graphene were obtained by comparing the predicted and measured contrasts. Roddaro et al. [17] imaged very thin graphite on a dielectric substrate using a normal optical microscope (Fig. 2.9). They also investigated the mechanism behind the strong visibility of graphite and discussed the importance of substrates and of the microscope objective used for the imaging. Jung et al. used a bilayer substrate comprising a thin dielectric layer on Si to identify and measure the effective optical properties of nanometer-thick graphene based materials [18].

Most recently, Huang et al. [19] studied bubbles in graphene by Raman spectroscopy and found that size-dependent oscillations in spectral intensity originate from optical standing waves formed in the vicinity of the graphene surface. Based on Raman data, the temperature distribution inside a graphene bubble was calculated which permitted the estimation of the thermal conductivity of graphene.

In fluorescence imaging light interference effects are used to determine the location of a fluorophore with nanometer precision [20, 21]. This technique named spec-

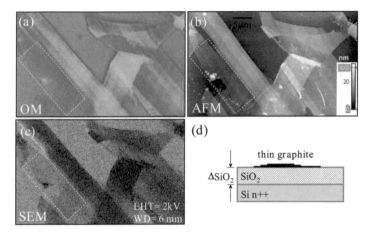

Fig. 2.9 Images of a large graphite flake containing regions of many different thicknesses. The pictures were taken using an optical microscope (**a**), an atomic force microscope (**b**), and a scanning electron microscope (**c**). An ultrathin graphite region (thickness < 2 nm) is highlighted by a dashed rectangle. **d** Graphite is deposited on top of a SiO_2/Si substrate with an oxide thickness SiO_2 of 500 nm. Reprinted with permission from reference [17], Copyright (2007) American Chemical Society

tral self-interference microscopy was inspired by the observation of Fromherz et al. [9] on the oscillation of fluorescent intensity as a function of the location or height of the fluorophore, in other words, its distance from a reflecting surface. By measuring the intensity of the fluorophore located at different known heights on a tailored SiO_2 substrate, the position of an unknown fluorophore can be determined with precision. When a fluorophore is present at larger distances (of the order of 10–20 λ) from a reflecting surface, light emitted from the fluorophore can undergo constructive and destructive interference multiple times even at the same height, leading to the formation of interference fringes in the emission spectrum. Spectral self-interference fluorescence microscopy (SSFM) utilizes this principle to reveal height information of the fluorophore.

Figure 2.10 shows the self-interference effect of the emission spectrum of fluorescein particles immobilized at different heights on a Si-SiO_2 substrate showing spectral oscillations. In sharp contrast is the spectrum of fluorescein immobilized on a glass substrate where there is no self-interference. Even very small height differences (10 nm in the figure) result in a visible shift in the fringes and changes the period of oscillation, although the latter effect is less apparent. By scanning the surface in a grid pattern with fluorescence emission spectra taken every 200 μm and followed by data processing, the axial positions of the fluorophores are determined at each spot and a surface map in 3D gray scale can be obtained at nanometer resolution. In practice, for fluorophores located at distances very close to each other ($< \lambda/2$), it is necessary to collect multiple spectra by independently scanning the excitation standing wave to deconvolute the contribution of the emitters present at different

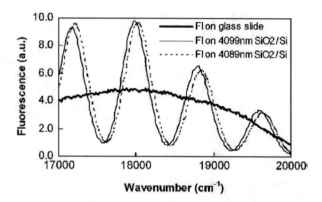

Fig. 2.10 Emission spectra of fluorescein immobilized on a glass slide and on top of a Si-SiO$_2$ chip with two different thicknesses of the oxide layer and surface profile reconstructed from self-interference spectra. Figure reproduced with permission from [20] SPIE copyright (2002)

heights to the total spectrum. More details on the theory, experimental set up and the reconstruction techniques employed are given in references [20, 21].

References

1. W.S. Bacsa, J.S. Lannin, Bilayer interference enhanced Raman spectroscopy. Appl. Phys. Lett. **61**, 19 (1992)
2. O. Wiener, Stehende Lichtwellen und die Schwingunsrichtung polariertes Lichtes. Ann. Phys. **276**(6), 203 (1890)
3. P. Ranson, R. Ouillon, J.-P. Pinan-Lucarré, A centenary, G. Lippmann, Nobel prize of Physics 1908 for colour photography. Europhys. News **39**(6), 18–22 (2008)
4. R.T. Holm, S.W. McKnight, E.D. Palik, W. Lukosz, Interference effects in luminescence studies of thin films. Appl. Opt. **21**(14), 2512 (1982)
5. J.W. Ager III, D.K. Veirs, G.W. Rosenblatt, Raman intensities and interfernce effects for thin films adsorbed on metals. J. Chem. Phys. **92**, 2067 (1990)
6. N. Umeda, Y. Hayashi, K. Nagai, A. Takayanagi, Scanning Wiener fringe microscopy with optical fiber tip. Appl. Opt. **31**, 4515 (1992)
7. W.S. Bacsa, A. Kulik, Interference scanning optical probe microscopy. Appl. Phys. Lett. **70**, 3507 (1997)
8. A. Kramer, T. Hartmann, R. Escher, R. Guckenberg, Scanning near-field fluorescence microscopy of thin organic films at the water/air interface. Ultramicroscopy **71**, 123 (1998)
9. A. Lambacher, P. Fromherz, Fluorescence interference-contrast microscopy on oxidized silicon using a monomolecular dye layer. Appl. Phys. A **63**, 207 (1996)
10. K.S. Novoselov, A.K. Geim, S.V. Morozov, D. Jiang, Y. Zhang, S.V. Dubonos, I.V. Grigorieva, A.A. Firsov, Electric field effect in atomically thin carbon films. Science **306**(5696), 666 (2004)
11. E. Hecht, *Optics*, 5th edn. (Pearson Education, London, 2016)
12. M.V. Klein, T.E. Furtak *Optics*, 2nd edn. (Wiley, Hoboken, 1986)
13. G.R. Fowles, *Introduction to Modern Optics*, 2nd edn. (Dover Publications INC., New York, 1968, 1975)

14. R. Carles, C. Farcau, C. Bonafos, G. Benassayag, M. Bayle, P. Benzo, J. Groenen, A. Zwick, Three dimensional design of silver nanoparticle assemblies embedded in dielectrics for Raman spectroscopy enhancement and dark-field Imaging. ACS Nano **5**(11), 8774 (2011)
15. J. Sukmanowski, C. Paulick, O. Sohr, K. Andert, F.X. Royer, Light absorption enhancement in thin silicon layers. J. Appl. Phys. **88**, 2484 (2000)
16. J. Kim, F. Kim, J. Huang, Seeing graphene-based sheets. Mater. Today **13**(3), 28 (2010)
17. S. Roddaro, P. Pingue, V. Piazza, V. Pellegrini, F. Beltram, The optical visibility of graphene: interference colors of ultrathin graphite on SiO2. Nano Lett. **7**, 2707 (2007)
18. I. Jung, M. Pelton, R. Piner, D.A. Dikin, S. Stankovich, S. Watcharotone, M. Hausner, R.S. Ruoff, Simple approach for high-contrast optical imaging and characterization of graphene-based sheets. Nano Lett. **7**, 3569 (2007)
19. Y. Huang, X. Wang, X. Zhang, X. Chen, B. Li, B. Wang, M. Huang, C. Zhu, X. Zhang, W.S. Bacsa, F. Ding, R.S. Ruoff, Raman spectral band oscillations in large graphene bubbles. Phys. Rev. Lett. **120**, 186104 (2018)
20. A.K. Swan, L. Moiseev, Y. Tong, S.H. Lipoff, W.C. Karl, B.B. Goldberg, M.S. Ünlü, High resolution spectral self-interference fluorescence microscopy, in Proceedings of SPIE, vol. 4621. Three-Dimensional and Multidimensional Microscopy: Image Acquisition and Processing IX. Accessed 15 May 2002
21. M.S. Ünlü, A. Yalcin, M. Dogan, L. Moiseev, A. Swan, B.B. Goldberg, C.R. Cantor, Applications of optical resonance to biological sensing and Imaging: I. Spectral self-interference microscopy, in *Biophotonics*, ed. by L. Pavesi, P.M. Fauchet (Springer, Berlin, 2008)

Chapter 3
Intermediate Field and a Single Point Scatterer on a Surface

3.1 Introduction

In the previous chapter, we saw how the formation of surface standing waves can be exploited to enhance the optical field on an interference substrate. Thus, interference substrates can be particularly useful for the observation of single atomic or molecular layers. Given that optical standing waves can be imaged using an optical fiber probe in collection mode [1], the question arises as to whether it is possible to use this technique to observe single nanoparticles. As we know, in lens based optical microscopy, the lateral resolution is limited by the size of the wavelength or, in other words, it is diffraction limited: the lateral resolution cannot be larger than $\lambda/2$.

In nearfield optics, the optical probe is scanned in close proximity to the surface ($<10\,\mathrm{nm}$) and the probe is controlled using force feedback, which means that optical resolution below the diffraction limit can be obtained. However, the optical resolution in nearfield optics is limited by the aperture size since smaller apertures result in greatly reduced light transmission and hence the aperture size needs to be in the $50 - 100\,\mathrm{nm}$ range, which is one order of magnitude larger than the distance of the probe to the substrate (around $10\,\mathrm{nm}$). This has the consequence that probe-induced effects are not negligible in nearfield optics.

In the following section we show how to use surface standing waves to move the optical fiber in a plane parallel to the surface at larger distances. Typically, when a surface is imaged using an optical probe, parallel fringes are observed due to the crossing of standing waves through the image plane. Surface standing waves are oriented parallel to the surface. From the spacing of these fringes, the tilt angle can be estimated and we can then correct the image plane of the piezo-electric displacement table. This means that images can be recorded in the proximity of the surface without any force feedback signal and at variable distances. Figure 3.1 shows an optical probe consisting of a pointed optical fiber with a metal coating on the side walls and an aperture at the tip. Light either scatters at the edge (I) or reflects off the surface (II) before propagating in the optical fiber to the photo detector. Scattering at the edge is proportional to the local field intensity while the light reflected off the surface

© The Author(s), under exclusive licence to Springer Nature Switzerland AG 2020
W. Bacsa et al., *Optics Near Surfaces and at the Nanometer Scale*,
SpringerBriefs in Physics, https://doi.org/10.1007/978-3-030-58983-7_3

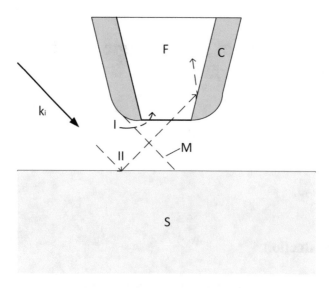

Fig. 3.1 Optical probe near substrate (S). The incident wave k_i illuminates the substrate and the optical probe on one side. The optical probe F is coated with a metal C. The probe detects the intensity at the aperture on the illuminated side at the edge (I) and through reflection on the substrate (II). A shadow of the standing wave is formed (M)

is constant and drops off monotonously at a distance smaller than the size of the aperture d_{ap}, that is

$$z < \frac{d_{ap}}{2\tan(\theta_i)} \tag{3.1}$$

where θ_i is the angle of incidence.

The aim of this chapter is to analyze the interference fringes recorded by an optical fibre when an incident wave interacts with a nanoparticle resulting in absorption and emission of photons, or simply scattering. The scattered wave interferes with the incoming wave with a phase which depends on the electronic resonance in the particle. We assume that a nanoparticle can be approximated as a single point scatterer and that the size of the nanoparticle is around two orders of magnitude smaller than the optical wavelength. We study here the interference pattern at variable distances between the particle and a plane parallel to the surface or a plane perpendicular to the surface (in fact the analysis is easily adapted to any angle) [2, 3].

It is interesting to note that although the nanoparticle cannot be directly viewed by optical waves, the interference pattern may be observed using an optical fibre from which the position of the particle may be deduced. In this chapter we will discuss the mathematics behind interference patterns and thereafter show how this technique may be applied in lensless imaging [4]. In the following section we first describe in more detail the characteristics of the optical field outside the near field region.

3.2 Inline Holography and the Intermediate Field

In holography, the wave field emerging from the object is recorded. This is achieved by generating an interference pattern formed by the wave emerging from the object and a reference wave. The reference wave is coherent with the incident wave illuminating the object, that is, the reference wave is a copy of the incident beam. Once this interference pattern is recorded, the object can be removed. When the recorded interference pattern is illuminated, the same wave field of the object or image is regenerated. This means the interference pattern acts as an optical grating and generates the wave field. In digital holography, the interference pattern is recorded on an array detector and the image is computer generated (numerical reconstruction). It is also possible to record holograms directly from the interference pattern of the object with the incident beam which is referred to as holography without a reference beam or inline holography. This technique has the advantage of increasing resolution due to the small distance between the image plane and the substrate. In a similar manner, standing wave patterns can be recorded on an array detector and the image can be generated by numerical reconstruction. In digital holographic microscopy, a point light source is used and the spreading of the light source leads to magnification of the image.

As already mentioned, the incident beam is scattered by the nanoparticle which is assumed to be a point scatterer. When considering the point scatterer as an electronic oscillator, the resulting scattered wave is an oscillating dipole wave whose electric field can be derived from the oscillating current induced by the incident field [5]. If the dipole is oriented along the z axis then the charge oscillates along z and the electric field only has two components, which are both in a plane of symmetry of the dipole,

$$E_r = \frac{2\cos(\theta)}{4\pi\epsilon_0}\left(\frac{p}{r^3} + \frac{\dot{p}}{r^2 c}\right), \qquad E_\theta = \frac{\sin(\theta)}{4\pi\epsilon_0}\left(\frac{p}{r^3} + \frac{\dot{p}}{r^2 c} + \frac{\ddot{p}}{r c^2}\right), \quad (3.2)$$

where p is the dipole moment of the nanoparticle and r is the distance between the probe and the nanoparticle

$$p = aq\cos(\omega t) = p_0 \cos(\omega t) \tag{3.3}$$

hence

$$\dot{p} = -p_0\omega\sin(\omega t), \qquad \ddot{p} = p_0\omega^2\cos(\omega t). \tag{3.4}$$

It is assumed that the distance r is substantially larger than the dimension of the dipole a.

Focusing on the E_θ expression (which is the more complex) and substituting for p we obtain

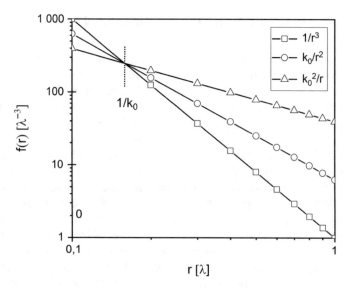

Fig. 3.2 The three components of the transverse field as a function of distance r. The $1/r$ component dominates at distances greater than $1/k_0$

$$E_\theta = \frac{p_0 \sin(\theta)}{4\pi \epsilon_0} \left(\frac{\cos(\omega t)}{r^3} - \frac{k_0 \sin(\omega t)}{r^2} - \frac{k_0^2 \cos(\omega t)}{r} \right) \qquad (3.5)$$

where $k_0 = \omega/c$. It is quite clear that, depending on the size of r, different components dominate. For very small r the first term, which is proportional to $1/r^3$, is obviously dominant whereas for large r, the final term dominates. In standard calculations only the final term is considered in the electric field, with the implicit assumption that

$$k_0^2 \gg \frac{k_0}{r} \gg \frac{1}{r^2} . \qquad (3.6)$$

The relative contribution of each term is shown in Fig. 3.2 (note the use of a logarithmic scale, the differences rapidly become large). The crossing point occurs at $r = 1/k_0 = \lambda/(2\pi) \approx \lambda/6$. Hence for distances $r \gg \lambda/6$ the dominant term in the electric field is

$$E_\theta = -\frac{p_0 \sin(\theta)}{4\pi \epsilon_0} \frac{k_0^2 \cos(\omega t)}{r} . \qquad (3.7)$$

Experimentally, this means that as long as the probe is further from the scatterer than $1/k \approx \lambda/6$ then the $1/r$ term dominates. For example, using a laser with emission at 400 nm this corresponds to a distance of 67 nm. This means that a lateral resolution of the order of 70 nm (or 100 nm when using a wavelength of 600 nm) can be expected when imaging right outside the near field region without any loss of resolution. That is the same lateral resolution obtained in near field optics.

For $r \gg \lambda/(2\pi)$ the radial component $E_r \propto 1/r^2$ is of the same magnitude as terms neglected in E_θ and so may also be neglected (apart from the fact that E_r is longitudinal and does not propagate into the far field).

If the dipole wave is a result of a propagating plane wave of the form (2.1) then a propagating dipole wave is generated

$$E_\theta = -\frac{p_0 \sin(\theta) k_0^2 \cos(\mathbf{k} \cdot \mathbf{r} - \omega t)}{r} . \tag{3.8}$$

The scattered field interferes with the incident beam and forms interference fringes. The interference term of two interfering waves (2.40) is

$$|\mathbf{E}_{m1}||\mathbf{E}_{m2}| \, \cos((\mathbf{k_1} - \mathbf{k_2}) \cdot \mathbf{r}) . \tag{3.9}$$

Since E_{m2} is proportional to $1/r$ the interference term is proportional to

$$\frac{\cos((\mathbf{k_1} - \mathbf{k_2}) \cdot \mathbf{r})}{r} \tag{3.10}$$

The other two terms of (2.40) are either constant or proportional to $1/r^2$. The term proportional to $1/r^2$ falls off faster with increasing r compared to the interference term and so is neglected in the following.

To understand the particularity of the intermediate field one needs to consider the components of the electric field in more detail. In the vicinity of the scatterer, $r \to 0$, all terms are large but with $1/r^3$ dominating. At large distances the influence of all terms is marginal. However when the distance is comparable to the size of the image, the $1/r$ factor varies strongly and significantly modifies the image contrast Fig. 3.3. Hence we define the **intermediate field** as the region where the $1/r$ term dominates and is still noticeable, specifically we define this as encompassing distances between $\lambda/6$ and 10λ.

3.3 Interference Fringes from a Single Point Scatterer

Interference of two waves, namely the incident plane wave and the dipole wave emerging from the point scatterer, give rise to interference fringes. The superposition of waves was discussed in the previous chapter where it was shown that the energy density or intensity contains the interference term

$$I_c = \cos(\phi_2 - \phi_1) = \cos\left((\mathbf{k_2} - \mathbf{k_1}) \cdot \mathbf{r} + \psi\right) , \tag{3.11}$$

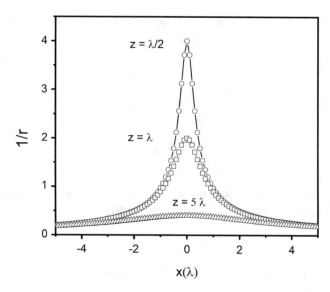

Fig. 3.3 Variation of the $1/r$ factor across the image 10λ in size when the image plane is at distance of $\lambda/2$, λ and 5λ. The function $1/r$ peaks when the image plane is at a distance smaller than the size of the image

see Eq. (2.40) (we have reintroduced the phase here which was previously omitted to shorten the algebra). Without loss of generality we may define the coordinate system such that the point scatterer is located at the origin of the (x, y, z) co-ordinate system and that the incident wave direction is contained in the $y - z$ plane and making an angle θ_i with the z-axis, $\mathbf{k}_1 = (0, \sin\theta_i, -\cos\theta_i)$. For a point source $\mathbf{k}_2$ is directed radially outwards, that is along $\mathbf{r}$, so $\mathbf{k}_2 = k_2 \cdot \hat{\mathbf{r}}$ and the position vector $\mathbf{r} = (x, y, z) = r(\sin\theta\cos\phi, \sin\theta\sin\phi, \cos\theta)$ (Fig. 3.3).

The two components required to calculate the interference term, I_c, are

$$\mathbf{k}_1 \cdot \mathbf{r} = kr(\sin\theta_i \sin\theta \sin\phi - \cos\theta_i \cos\theta), \qquad \mathbf{k}_2 \cdot \mathbf{r} = kr. \qquad (3.12)$$

Maxima of the interference patterns occur when the argument of (3.11) is some even multiple of π,

$$kr(1 - \sin\theta_i \sin\theta \sin\phi + \cos\theta_i \cos\theta) = \pm 2n\pi - \psi. \qquad (3.13)$$

To decide which sign to take on the right hand side, consider the region where $\phi = 0$ (this corresponds to the plane $y = 0$). The left hand side of (3.13) reduces to $kr(1 + \cos\theta_i \cos\theta)$. The product $-1 \leq \cos\theta_i \cos\theta \leq 1$ so the left hand side is non-negative. If Eq. (3.13) is to hold for any ψ, where $0 \leq \psi \leq 2\pi$, then it is clear that we must take the +ve branch, so

$$r(1 - \sin \theta_i \sin \theta \sin \phi + \cos \theta_i \cos \theta) = \left(n - \frac{\psi}{2\pi}\right)\lambda , \tag{3.14}$$

where $\lambda = 2\pi/k$ is the wavelength and $n \geq 1$. Since our experiment is in air the index of refraction has been set to unity, hence we have set $k_0 = k$. Note, if we take the - sign then this corresponds to theoretical fringes below $z = 0$. This could be of interest if the laser is on one side of a translucent surface and the measuring device on the other side. Equation (3.14) may also be written in Cartesian form

$$\sqrt{x^2 + y^2 + z^2} - y \sin \theta_i + z \cos \theta_i = \left(n - \frac{\psi}{2\pi}\right)\lambda . \tag{3.15}$$

Equations (3.14) and (3.15) define a family of three-dimensional surfaces. We will see later that they are paraboloid surfaces showing where the maximum intensity occurs for different values of integer n. The surfaces get larger with increasing n. Once ψ is known these surfaces may be immediately plotted for each n. In practice when imaging the interference pattern the image plane cuts through these paraboloid surfaces. This means, for example, that with a horizontal image plane a set of ellipses should be observed, whereas for a vertical image plane parabolas will be observed.

For brevity, from now on, we denote

$$s_i = \sin \theta_i , \quad c_i = \cos \theta_i , \quad \mu_n = \left(n - \frac{\psi}{2\pi}\right)\lambda , \quad \Lambda_n = \mu_n - z_c c_i , \tag{3.16}$$

and work with the Cartesian description. If the image plane is horizontal (i.e. aligned with the plane $z = 0$) and at a distance $z = z_c$ from the surface, then Eq. (3.15) shows that

$$\sqrt{x^2 + y^2 + z_c^2} - y s_i + z_c c_i = \mu_n . \tag{3.17}$$

Taking $y s_i, z_c c_i$ to the right hand side and then squaring both sides of the equation

$$x^2 + y^2 + z_c^2 = \Lambda_n^2 + 2\Lambda_n y s_i + y^2 s_i^2 . \tag{3.18}$$

This may be rearranged into the form

$$\frac{x^2}{a_n^2} + \frac{(y - y_{cn})^2}{b_n^2} = 1 , \tag{3.19}$$

which is the equation for an ellipse with the centre located at $(0, y_{cn})$, where

$$y_{cn} = \frac{\Lambda_n \sin \theta_i}{\cos^2 \theta_i} , \tag{3.20}$$

and with minor and major axes

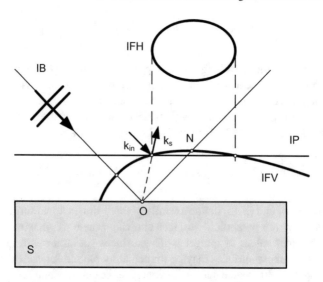

Fig. 3.4 Schematic of experimental setup: incident beam (IB), scatterer (O), substrate (S), image plane (IP), Vertical Interference Fringe (IFV, which has a parabolic form) and horizontal interference fringe (IFH, elliptical form). For simplicity only one interference fringe is shown. $\mathbf{k}_{in}$ and $\mathbf{k}_s$ are the wave vectors of the incident and scattered wave. The line through ON shows the reflected beam direction. The dashed line indicates the angular range of $\mathbf{k}_s$

$$a_n = b_n \cos \theta_i , \qquad b_n = \frac{\sqrt{\Lambda_n^2 - z_c^2 \cos^2 \theta_i}}{\cos^2 \theta_i} . \qquad (3.21)$$

Solutions therefore depend on the angle of incidence, distance between the image plane and surface z_c and wavelength. The first of Eqs. (3.21) shows the interesting result that the incident angle can be determined simply by taking the ratio of the principal axis of the fringes. Note, if we apply the same procedure on a vertical plane, for example by choosing $x = x_c$ instead of $z = z_c$, then we end up with an implicit equation for a parabola (this is discussed in Sect. 3.5).

An interesting feature of these equations is that, provided the incident angle is known, a_n, b_n only involve two unknowns, Λ_n, z_c (or μ_n, z_c). Consequently we may determine their values by taking a minimum of two measurements.

We will now illustrate the method via a particular experiment where a Si nanoparticle is observed on a Si surface. Figure 3.4 depicts the experimental set-up. The sample is illuminated with a laser (of wavelength $\lambda = 632$ nm) at an estimated incident angle ($\theta_i \approx 53° \approx 0.925$ radians). The local intensity is recorded by an optical fibre in collection mode. In Fig. 3.5 a horizontal interference pattern recorded by the optical fibre is shown. Note, the intensity of the incident beam, E_1^2, has been subtracted from the image while that of the nanoparticle, E_2^2, is negligible. The image was taken at an unknown height above the substrate, defined as $z = z_c$, the size of the image is $60 \times 60 \,\mu$m. The central series of ellipses is due to a single isolated nanoparticle, however we can also see some interaction with interference fringes

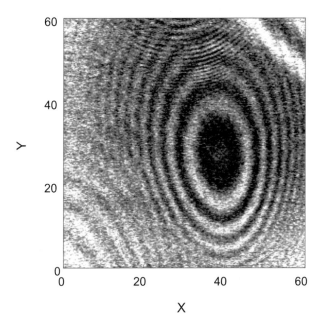

Fig. 3.5 Interference pattern due to interference of light scattered by a nanoparticle with that of a laser of wavelength 632 nm. The image size is $60 \times 60 \, \mu m$

from neighboring particles near the top right, top middle and bottom left of the image. The interference pattern shown corresponds to the interference between the incident beam and the light from the nanoparticle. A separate set of elliptical fringes will occur due to the reflected beam but this is outside of the limits of the figure. These will be larger than those generated by the incident beam and less intense.

By taking measurements from the interference pattern of Fig. 3.5 and comparing with Eqs. (3.19) and (3.21) we can determine the parameters necessary to characterise the pattern.

However, before we go further we note that most of the information taken from the figure does not correspond to the (x, y, z) co-ordinate system, which has its origin at the, as yet unknown, position of the nanoparticle. To deal with this we now introduce a second co-ordinate system $\mathbf{X} = (X, Y, Z)$ such that the origin is located at the bottom left corner of Fig. 3.5 with X horizontal, Y vertical and Z equivalent to z.

In Table 3.1 we give the measured values of the major and minor axis lengths as well as the position of the centre (in the (X, Y) system) for the first three brightest fringes of Fig. 3.5. These values were obtained by estimating the central part of the fringes at the extreme points of the ellipse, i.e. the highest and lowest points (to obtain b_n) and the most left and right points (to obtain a_n). Knowing the extreme points the centre is simply midway between them. In fact to reduce the effect of measurement

Table 3.1 Length of minor and major axes and position of centre of the first three fringes. All lengths are in μm

	a	b	(X_c, Y_c)
Fringe 1	7.21	12.25	(38.56, 26.96)
Fringe 2	11.49	19.21	(38.07, 27.71)
Fringe 3	14.60	24.44	(37.78, 28.29)

error we determine Y_c as the average of the Y values between the top and bottom points and X_c as the average of the left and right X values.

Below we will work with the first two fringes, since these are the clearest in the figure. In [4] more fringes are used and then an average taken. Taking a, b from the first two fringes in Table 3.1 and calculating $\theta_i = \cos^{-1}(a/b)$ indicates an average value $\theta_i = 0.934$ rads, which is very close to the quoted value of 0.925. Now, referring to Table 3.1 we denote the first semi-major axis $b_n = 12.25$ and the next as $b_{n+1} = 19.21$. The second of Eqs. (3.21) defines b_n in terms of the system parameters, Λ, z_c, θ_i. Noting that $\Lambda_{n+1} = \Lambda_n + \lambda$ we may write

$$b_n = 12.25 = \frac{\sqrt{\Lambda_n^2 - z_c^2 \cos^2 \theta_i}}{\cos^2 \theta_i} \tag{3.22}$$

$$b_{n+1} = 19.21 = \frac{\sqrt{\Lambda_{n+1}^2 - z_c^2 \cos^2 \theta_i}}{\cos^2 \theta_i} = \frac{\sqrt{(\Lambda_n + \lambda)^2 - z_c^2 \cos^2 \theta_i}}{\cos^2 \theta_i} . \tag{3.23}$$

Since λ, θ_i are known these provide two simultaneous equations for the two unknowns Λ_n, z_c which may easily be solved

$$\Lambda_n = \frac{(b_{n+1}^2 - b_n^2) \cos^4 \theta_i - \lambda^2}{2\lambda}, \qquad z_c = \sqrt{\frac{\Lambda_n^2 - b_n^2 \cos^4 \theta_i}{\cos^2 \theta_i}} . \tag{3.24}$$

In this case we find $z_c = 35.16\,\mu$m, $\Lambda_n = 21.35\,\mu$m, which then indicates $\mu_n = 42.25\,\mu$m. Now, we note that

$$\mu_n = \left(n - \frac{\psi}{2\pi}\right)\lambda \quad \Rightarrow \quad n - \frac{\psi}{2\pi} = 66.85 . \tag{3.25}$$

Since n is an integer and $\psi < 2\pi$ we immediately deduce $n = 67$, $\psi = 2\pi \times (67 - 66.85) \approx 0.898$ radians.

With the values of z_c, n, ψ known we are now able to reproduce every ellipse on the figure by varying $n \geq 67$. In fact, with just the value of ψ we could use Eq. (3.15) to find all surfaces of maximum intensity, below or above the image plane, with $n \geq 1$. Further, as n increases the value of ψ becomes irrelevant to the interference pattern (see the later discussion on sensitivity). However, at present the surfaces are

Fig. 3.6 Rotated coordinate and displaced system

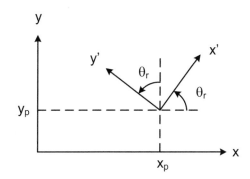

only described in the **x** co-ordinate system, whereas the image is in the **X** system. To be of any practical use we must be able to translate the theoretical results to the image plane.

3.4 Conversion Between Co-ordinate Systems

To compare with the experimental image and also to determine the position of the particle in the image plane requires converting between the two co-ordinate systems. The **x** system has been chosen so that the particle is located at the origin with the y-axis aligned along the major axis of the brightest ellipse. In Fig. 3.5 the ellipses are far from the origin and the major axis is not parallel to the y-axis. The conversion therefore requires a rotation and translation.

To obtain the values of Table 3.1 we recorded the top and bottom points of the first two fringes (in the **X** system). This permits us to calculate the rotation angle between the x and X axes. Since the Y axis is vertical the tangent of the rotation angle is simply the distance between the X co-ordinates divided by the distance between the Y co-ordinates of any of the ellipses, $\tan \theta_r = (X_t - X_b)/(Y_t - Y_b)$ where t signifies top and b bottom. From this we obtain $\theta_r \approx -4.30°$. The negative sign indicates that in moving from the **x** to the **X** co-ordinate system we move 4.30° in the clockwise direction (Fig. 3.6).

Combining the standard rotation matrix with a translation we may now relate the two co-ordinate systems,

$$
\begin{pmatrix} X \\ Y \\ Z \end{pmatrix} = \begin{pmatrix} X_p \\ Y_p \\ 0 \end{pmatrix} + \begin{pmatrix} \cos \theta_r & \sin \theta_r & 0 \\ -\sin \theta_r & \cos \theta_r & 0 \\ 0 & 0 & 1 \end{pmatrix} \begin{pmatrix} x \\ y \\ z \end{pmatrix} , \tag{3.26}
$$

where $(X_p, Y_p, 0)$ is the unknown position of the particle in the **X** system (i.e. the **X** co-ordinate when $\mathbf{x} = \mathbf{0}$).

To determine the position X_p it is sufficient to equate two known points in the two systems. For example, according to Eq. (3.20) the centre of the first visible fringe, corresponding to $n = 67$, is at

$$y_{c(67)} = (\mu_{67} - z_c \cos \theta_i) \frac{\sin \theta_i}{\cos^2 \theta_i} \approx 48.55 . \tag{3.27}$$

The position $(x_{c(67)}, y_{c(67)}) = (0, 48.55)$ must coincide with the centre quoted in Table 3.1, $(X_{c(67)}, Y_{c(67)}) = (38.56, 26.96)$. From the transform (3.26) we have

$$(X_p, Y_p) = (X_{c(67)} - y_{c(67)} \sin \theta_r, Y_{c(67)} - y_{c(67)} \cos \theta_r)$$
$$\approx (42.09, -21.45) . \tag{3.28}$$

We have found the position of our nanoparticle and with only measuring two distances!

A key part of practical studies such as this is to verify the theory via the experimental data. Firstly, substituting the determined values of z_c, n, ψ into the theoretical definitions of a_n and b_n, Eq. (3.21), the predicted axis lengths are

$$a_{67} = 7.28 \quad a_{68} = 11.42 \quad a_{69} = 14.49$$
$$b_{67} = 12.25 \quad b_{68} = 19.21 \quad b_{69} = 24.38 .$$

These are all within 1% of the experimental values quoted in Table 3.1. Another obvious test is to compare the theoretical fringes with the experimental ones. Using Eq. (3.19) we may obtain the ellipses in the x, y system (note x varies between $[-a_n, a_n]$ and $y - y_c \in [-b_n, b_n]$). The transformation (3.26) then determines the ellipses in the X system. The results are shown in Fig. 3.7 where we also show the position of the nanoparticle predicted by Eq. (3.28). The agreement is remarkably good given the simplicity of the calculation.

3.4.1   Accuracy of the Results

A significant advantage of finding an analytical solution is that its accuracy is typically of the order of any approximations made: if we know the order of the approximations then we know the order of the errors. In the present work we have obtained exact solutions and so the sole source of error comes from the measurements. In [4] to reduce errors a number of fringe measurements were used to calculate z_c, Λ_n and then an average was taken. Here we simply used two values, b_n, b_{n+1}. If these are out by say 1% then there will be an error of the order of 1% in z_c, Λ_n, μ_n. However, if we consider the value of

$$\mu_n = \Lambda_n + z_c c_i = n\lambda \left(1 - \frac{\psi}{2n\pi}\right) , \tag{3.29}$$

Fig. 3.7 Comparison of theoretical curves with experimental image. Also shown is the position of the particle, (42.09, −21.45), causing the interference

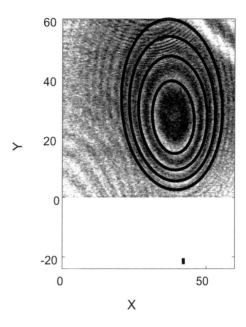

then it is clear that for large n the phase shift plays a small role. In the present case where $n = 67$ we find $\psi/(2n\pi) \sim 1/n \sim 0.015$: so it contributes around 1.5% of μ_n. Consequently, an error in μ_n of the order of 1% is of the same order as the total contribution of ψ, meaning that the prediction of ψ could be out by 100%! Our method is therefore very accurate for determining the position of the particle and reproducing interference patterns, but may be highly inaccurate at predicting the phase shift. In fact the phase shift is notoriously difficult to measure. No doubt this sensitivity is a reason for the difficulty in predicting the phase shift using other methods and the reason that it is often the subject of complex retrieval methods [6].

3.5 Vertical Cross-Section

The previous analysis was focussed on replicating a horizontal interference pattern and so locating a nanoparticle in a specific experiment. However, the theory developed leading up to this is much more general. In this section we will demonstrate how the results may be used for different forms of pattern as well as an example where the axis lengths are not obvious.

In Sect. 3.3 we derived the equation for the three dimensional variation of the time-averaged energy. By choosing the argument of the cosine to be positive integer multiples of π we restricted the solutions to the surfaces of maximum energy. Then by fixing $z = z_c$ to be constant we confined the solutions further to be a cross-section on the z_c plane, resulting in a series of distinct ellipses. Obviously this was motivated by the experimental image but the analysis could equally well have been applied to different types of images, for example with a vertical cross-section.

The starting point for the previous analysis was the equation

$$k\left(\sqrt{x^2 + y^2 + z^2} - y\sin\theta_i + z\cos\theta_i\right) = 2n\pi - \psi , \qquad (3.30)$$

with $2n$ an even integer. The choice of $2n$ was to make the cosine of the left hand side a maximum and so correspond to the brightest parts of the graph. However we could take any other value, not necessarily integer, to find different surfaces. The obvious one being the minima, i.e. the darkest parts of the interference pattern, which correspond to $2n + 1$ (again with integer n).

To illustrate this discussion we now examine a vertical cross-section in a constant x-plane, $x = x_c$, and plot the curves of minimum brightness. The curves of minimum brightness occur when $I_c = -1$ so are described by

$$k\left(\sqrt{x_c^2 + y^2 + z^2} - y\sin\theta_i + z\cos\theta_i\right) = (2n + 1)\pi - \psi , \qquad (3.31)$$

with $n \geq 1$ an integer. This leads to

$$x_c^2 + y^2 + z^2 = \left(\mu_n' + y\sin\theta_i - z\cos\theta_i\right)^2 , \qquad (3.32)$$

where $\mu_n' = ((n + 1/2) - \psi/(2\pi))\lambda$. Expanding the right hand side and rearranging we find

$$c_i^2 y^2 + s_i^2 z^2 - 2\mu_n' s_i y + 2\mu_n' c_i z + 2s_i c_i yz = \Lambda_n'^2 , \qquad (3.33)$$

where $\Lambda_n'^2 = \mu_n'^2 - x_c^2$. Although not obvious, this is an implicit equation for a parabola. To find a simpler form we need to work in a different co-ordinate system where the parabola takes on the standard form

$$Az' = By'^2 + Cy' + D . \qquad (3.34)$$

For this we define the new system

$$\begin{pmatrix} y' \\ z' \end{pmatrix} = \begin{pmatrix} \cos\alpha & -\sin\alpha \\ \sin\alpha & \cos\alpha \end{pmatrix} \begin{pmatrix} y \\ z \end{pmatrix} \qquad (3.35)$$

for some unknown angle α. The matrix acts to rotate the axes α radians in a counter-clockwise direction. Replacing y', z' in the parabola equation leads to

$$B(c_\alpha^2 y^2 + s_\alpha^2 z^2) + (Cc_\alpha - As_\alpha)y - (Cs_\alpha + Ac_\alpha)z - 2Bc_\alpha s_\alpha yz = -D . \qquad (3.36)$$

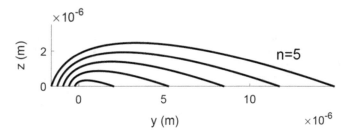

Fig. 3.8 Parabolas depicting the position of the first five fringes of minimum brightness through the plane $x = 0$, corresponding to $n = 1, 2, 3, 4, 5$, with $\psi = 0.898$, $\theta_i = 0.934$ radians

Equations (3.33) and (3.36) are equivalent provided $A = -2\mu'_n$, $B = 1$, $C = 0$, $D = -\Lambda'^2_n$ and $\alpha = -\theta_i$ then we find the parabola equation

$$z' = \frac{\Lambda'^2_n - y'^2}{\mu'_n} .\qquad(3.37)$$

We may convert back to the original system by inverting (3.35)

$$\begin{pmatrix} y \\ z \end{pmatrix} = \begin{pmatrix} c_i & -s_i \\ s_i & c_i \end{pmatrix} \begin{pmatrix} y' \\ z' \end{pmatrix}.\qquad(3.38)$$

This indicates that the $y' - z'$ co-ordinate system results from rotating $y - z$ by θ_i radians in the clockwise direction (which we may have expected from physical considerations) and so y' points in the line of the incident wave.

Taking the previous values of $\psi = 0.898$, $\theta_i = 0.934$ radians we obtain curves of the form shown in Fig. 3.8. These correspond to the first five curves of minimum brightness which would be observed if a vertical image was obtained in the previous experiment.

3.6 Numerical Image Reconstruction

Fitting of the elliptical fringes to experimental measurements as described in the previous chapter allows us to find the angle of incidence θ_i, the distance between the image plane and the substrate z_c as well as the location of the nanoparticle. Knowing z_c is useful when scanning the probe in the proximity of the surface. Since no direct interaction field is used to control the probe, apart from the surface standing waves to orient the image plane parallel to the surface, analyzing the fringe spacing of a single scatterer allows us to find the distance between the image plane and surface.

Knowing θ_i and z_c is also useful in reconstructing the image from the diffraction images. There are two possibilities to do this, either by correlating the image as

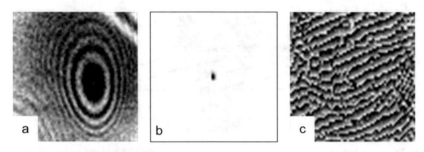

Fig. 3.9 Experimental image **a** with reduced resolution (64×64), **b** reconstructed image amplitude and **c** phase. Image size is $60 \times 60\,\mu m$

obtained with the model with the experimental image or by calculating the Fresnel integral. Calculating the Fresnel integral means that each image point is treated as a source of a spherical wave with an amplitude which corresponds to the intensity of the image point by taking into account the phase from the angle of incidence. The superposition of the individual waves is then calculated in the plane of the substrate. For each image point in the plane of the substrate $\mathbf{rs_{p,q}}$ one needs to sum over all the spherical waves $\mathbf{r_{i,j}}$ emanating from the recorded image.

$$I O_{p,q} = \sum_{i,j}^{N} I m_{i,j} \left(\frac{i \ (k_0 \mid \mathbf{r}_{i,j} - \mathbf{rs}_{p,q} \mid + \mathbf{k}_i \cdot \mathbf{r}_{i,j})}{\mid r_{i,j} - r s_{p,q} \mid} \right) \tag{3.39}$$

where

$$\mathbf{r}_{i,j} := (i \, \Delta, \, j \, \Delta, \, h), \quad \mathbf{rs}_{p,q} := (p \, \Delta, \, q \, \Delta, \, 0) \tag{3.40}$$

and

$$\mathbf{k}_i = k_0(-\sin(\theta)\cos(\phi), \, -\sin(\theta)\sin(\phi), \, -\cos(\theta)), \tag{3.41}$$

$k_0 = 2\pi/\lambda$ and ϕ is the angle of the incident beam.

Figure 3.9 shows different images of a single particle but with (a) a reduced resolution to speed up the calculation of the Fresnel integral, (b) the result of the evaluation of the Fresnel integral, the square of the amplitude and (c) the phase of $\mathbf{rs}$. It may be observed that the linear superposition of the individual spherical waves converge to a single point, the point scatterer which caused the interference pattern. The half width of the peak at the location of the point scatterer is 1.9λ when recording the image at a distance of $z_c = 57.06\lambda$. To increase the image resolution the distance z_c needs to be reduced. We note that increasing the number of points in the recorded image does not increase resolution in the reconstructed image.

3.7 At Distances Smaller Than Half the Wavelength

When recording images at distances smaller than half the wavelength of the incident beam the distances are too small to notice any effect of diffraction and the image resembles the direct image. The advantage here is that the image plane may be adjusted parallel to the surface using surface standing waves with no feedback signal to control the probe in the proximity of the surface. The estimated image resolution is 35 nm or 1/18 of the incident beam wavelength [7]. This spatial resolution is comparable to what is observed in nearfield optics. Interestingly the spatial resolution is higher than $\lambda/2\pi$ which is attributed to the fact that the terms in the electrical field proportional to $1/r^2$ and $1/r^3$ contribute to increasing the field at the location of the scatterer.

3.8 Conclusions

Considering the interference of a plane wave with a spherical wave, the fringe patterns in the horizontal and vertical planes with respect to the support surface were derived. The fringes in the horizontal plane are elliptical and in the vertical plane parabolic. In the horizontal plane the elliptical fringes depend on z_c the distance between the image and the object plane. By comparing the analytic expressions for the fringes with experimental images permits the determination of z_c which is important when scanning an optical probe in the proximity of the surface. In addition the angle and direction of the incident beam and location of the point scatterer were calculated. Furthermore it was found that the influence of the phase of the scattered wave on the interference fringes is reduced when increasing the distance from the point scatterer. The reconstruction can be obtained by evaluation of the Fresnel integral using the z_c determined from the fringe pattern. High lateral resolution is obtained at smaller z_c (Fig. 3.10).

Fig. 3.10 Recorded image in collection mode in the proximity of the Ag island film on a interference substrate. Estimated image resolution 35 nm. Reprinted with permission from reference [7], Copyright (1997) American Institute of Physics

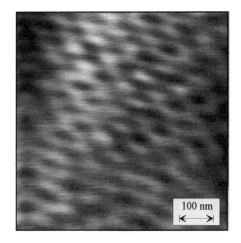

All the calculations have been done here assuming that the term proportional to $1/r$ is dominant. At distances smaller than half the wavelength the terms proportional $1/r^2$ and $1/r^3$ also contribute. This has the consequence that the intensity is increasingly localized improving spatial resolution.

It was further assumed that the dipoles of the substrate do not contribute in the image recorded in a plane parallel to the surface of the substrate. The response of the medium to the incident wave generates a reflected wave which will be diffracted at the point scatterer. A closer examination shows that the diffraction fringes in this case are larger and do not contribute to the first few reflected fringes and have therefore been neglected here.

References

1. N. Umeda, Y. Hayashi, K. Nagai, A. Takayanagi, Scanning Wiener-fringe microscope with an optical fiber tip. Appl. Opt. **31**, 4515 (1992)
2. B. Levine, A. Kulik, W.S. Bacsa, Optical space and time coherence near surfaces. Phys. Rev. B **66**, 233404 (2002)
3. W. Bacsa, B. Levine, M. Caumont, B. Dwir, Local optical field variation in the neighborhood of a semiconductor micro-grating. J. Opt. Soc. Am. (JOSA) B **23**, 893 (2006)
4. T.G. Myers, H. Ribera, W.S. Bacsa, Optical diffraction from isolated nanoparticles (2019), arXiv:1902.08680v1
5. E. Hecht, *Optics*, 5th edn. (Pearson Education, London, 2016)
6. Y. Shechtman, Y.C. Eldar, O. Cohen, H.N. Chapman, J. Miao, M. Segev, Phase retrieval with application to optical imaging: a contemporary overview. IEEE Signal Process. Mag. 87 (2015)
7. W.S. Bacsa, A. Kulik, Interference scanning optical probe microscopy. Appl. Phys. Lett. **70**, 3507 (1997)

Chapter 4
Spectral Shifts from Nano-Emitters and Finite Size Effects of the Focal Spot

4.1 Consequence of the Nanoparticle Being Smaller Than the Focal Spot

Optical microscopy and optical spectroscopy are essential tools to study materials at small scales. Both techniques are non-invasive and work at ambient atmosphere which makes them extremely versatile. However, in practice, the resolution of optical microscopes, including conventional widefield, confocal, and two-photon instruments is limited by several fundamental physical factors. For example, in conventional optical microscopy, the resolution is limited by the size of the focal spot, which is directly related to the optical wavelength. Since the wavelength of visible light ranges between 400 and 700 nm, the focal spot is usually larger than nanoparticles which are only a few nanometers in size. Thus, the measured optical response from the sample is averaged over an ensemble of nano objects. While such information can be very useful to demonstrate the overall behavior of nanoscale materials, individual particles cannot be characterised.

Optical spectroscopy of individual nanoparticles (e.g., semiconductor nanocrystals, nanotubes, nanowires, defects, dopants centers, etc.) is important both from the point of view of fundamental studies and for nanotechnology applications. For example, measurements on individual nano-emitters is important for carbon nanotubes where the electronic structure of an individual nanotube depends on its diameter. In typical macroscopic samples, different diameters are present and the obtained optical signal is averaged over many tubes of different diameters. In such cases, a way out is to use a highly diluted sample so as to have a single nanoparticle in the field of view of the microscope. Thus, spectroscopy of individual nanoparticles can be achieved by diluting the number of emitters in the focal spot to less than one. Recent progress in light detection techniques have thus made it possible to measure light emission/absorption from individual nano-emitters and obtain the corresponding spectroscopic information. The spectroscopic signature of the individual emitter, while being determined principally by its chemical composition, also depends to

W. Bacsa et al., *Optics Near Surfaces and at the Nanometer Scale*, SpringerBriefs in Physics, https://doi.org/10.1007/978-3-030-58983-7_4

Fig. 4.1 Nano emitter
displaced by the distance D
in the focal spot FS. Rays
emerging from the nano
emitter make an angle φ.
This causes a shift on the
array detector in the spectral
direction causing an apparent
spectral shift

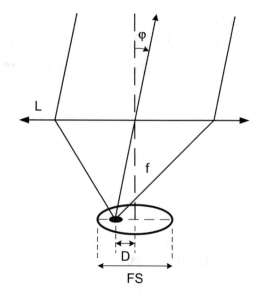

some extent on the shape and size of the nanoparticles, as well as its interaction with its immediate environment. The optical signal from a nanoparticle is also affected by applied fields such as electric and magnetic fields, or external parameters such as temperature and strain.

In spectroscopy, incident light from a light source is transmitted, absorbed or reflected through a sample. The outgoing light intensity emitted by a substance is measured over a broad spectral range by a spectrometer. When collecting optical spectra from nanoscale objects, the object plane of a microscope is imaged to the entrance slit of the spectrometer and the entrance slit is in turn imaged on the spectrometer array detector. The spectrometer grating has a given orientation and diffracts light of a given wavelength on the focal spot of the detector. The spectrometer grating thus disperses the light in a given direction which we will call the "spectral direction" which is perpendicular to the grating grooves. In the following, we show how the light emitted by nanoparticles smaller than the focal spot is analyzed by the spectrometer.

Let us consider light emanating from a nano emitter as shown in Fig. 4.1. Since the size of this nano emitter is smaller than the size of the focal spot, rays emitted by the nanoparticle emanate only from a fraction of the focal spot. A nano emitter displaced from the optical axis D but still within the focal spot FS emits rays which are tilted with respect to the optical axis by the angle φ.

This angle though small, causes a measurable shift in the image on the array detector. If the displacement of the image is in the spectral direction, an apparent spectral shift is recorded, which in reality, is due to the off-axis position of the nano-emitter. Since emission occurs only from a fraction of the focal spot of the microscope, only a fraction of the image of the focal spot on the array detector is illuminated. If the displacement is perpendicular to the spectral direction, there is no change in the observed spectral position.

It follows that even though a parallel beam can be focused to form a diffraction limited spot, an emitter that is smaller or has the same size as the focal spot does not lead to a parallel beam but to a slightly divergent beam. This divergence is obviously related to the size of the focal spot. The grating equation which is derived from the path length difference as a function of the angle of incidence and the wave length is given by

$$d(\sin \vartheta_i + \sin \vartheta_m) = m\lambda , \qquad (4.1)$$

where m is an integer, d the grating spacing, ϑ_i the angle of the incident beam and ϑ_m the angle of the diffracted beam.

A change in the angle of incidence thus has a direct effect on the diffraction angle which causes the apparent spectral shift in the spectral direction. This apparent spectral shift is entirely due to off axis emission and does not arise from changes in the emission wavelength of the emitter [1]. However, the apparent spectral shift has the effect that the beam is mapped on a different area on the detector within the image of the focal spot, even though the same wavelength λ_0 is emitted. In other words, a displacement in the object plane δx_{obj} is imaged to a displacement at the slit, $\delta x_{slit} = \delta x_{obj} M_{OS}$ [1]. The displacement in the object plane and displacement on the array detector are related and we can calculate the expected shift in spectral position $\delta \sigma$ from the displacement δx_{obj} of the nano emitter from the optical axis. Conversely, the observed spectral shift can be used to localize the nano emitter within the focal spot. Figure 4.2 shows how the two halves of the focal spot are imaged on different locations on the array detector in the spectral directions leading to an apparent spectral shift.

It is interesting to note that similar shifts have been observed in thermal infrared spectra of planetary surfaces where surface roughness strongly altered the slope of a thermal infrared spectrum, especially for large solar incidence angles. Roughness induced spectral slopes of 5% between 5 and 8 μm for regions of moderate topography were observed for the Moon surface. It was found that this shift was minimized by viewing near zero phase angle. By comparing the shifts obtained at different viewing angles, surface roughness and topology of the planetary surface could be determined [2].

4.2 Quantifying Apparent Spectral Shifts

Below, we demonstrate the finite size effect of the focal spot by considering a one dimensional nanoparticle such as a carbon nanotube. Carbon nanotubes can be one nanometer wide (single wall carbon nanotubes or SWNT) and several micrometers long. SWNTs can be grown across trenches on a silicon wafer in a direction perpendicular to the microfabricated trenches and studied individually without any substrate interaction. Moreover, their location can be verified by scanning electron microscopy and the substrate with SWNTs can then be displaced on the nanometer scale through

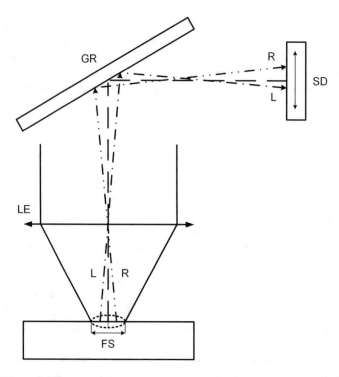

Fig. 4.2 Schematic of how the two halves of the focal spot (FS) are imaged (lens system LE) on the detector (SD). Left (L) and right (R) halves of the focal spot are imaged at different locations on the array detector causing an apparent spectral shift in the spectral direction (SD). This occurs only when the focal spot is not fully illuminated and for variations at scales smaller than the size of the focal spot

the focal point with a piezoelectric displacement table. Even though the spectral signal from an individual nanotube will be small, the signal can be enhanced by tuning the excitation wavelength to the electronic transition energy of the CNT, which has made possible the observation of individual nanotubes by optical methods. Figure 4.3 shows the spectral position of one of the in plane optical phonon modes (G+ Raman mode) of a carbon nanotube as a function of its position with respect to the optical axis of the microscope. The measurements were performed with two optical objectives with different magnifications with and without a nanotube. The measurements in the presence of an individual nanotube show a linear dependence of the spectral position with respect to the optical axis of the microscope while the measurement on silicon shows no spectral shift with displacement. This simple experiment demonstrates how a uniformly illuminated focal spot leads to no spectral shifts whereas when the focal spot is only partially illuminated, a spectral shift is observed. The factor of 2 higher spectral shift for one objective (x100) correlates with the 2 times higher magnification of the other objective (x50) [1].

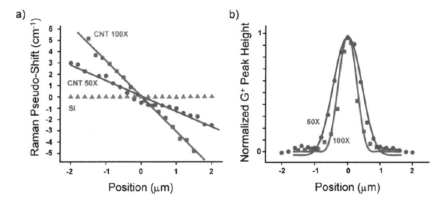

Fig. 4.3 Apparent spectral shift for a single nanotube (Raman G band) and LO optical phonon of silicon substrate as a function of position within the focal spot (objective 100×/0.9 NA square marks, 50×/0.5 NA objective, circular marks). Horizontal line (triangles) Si substrate (**a**) and G band intensity as a function of position of the nanotube compared with (**b**). Reprinted with permission from reference [1], Copyright (2008) American Chemical Society

The maximum measured shift in wavenumber, $5\,\mathrm{cm}^{-1}$, corresponds to a 1.5 µm displacement while a displacement within the FWHM of the focal spot intensity (280 nm) gives a shift of $\sim 2\,\mathrm{cm}^{-1}$. These values are in accordance with the variation in the reported values of Raman G+ band frequency found in the literature. It is noted here that the G+ band frequency is an important measured parameter to structurally characterize a nanotube and is critically influenced by strain, temperature, and charge transfer. Thus, the apparent spectral shift though small, can be confused with strain, temperature or charge transfer variations and this has to be taken into account when imaging nano emitters smaller than the focal spot.

The dispersion can be calculated using

$$\Delta\sigma = \Delta x_{obj} M_{OS} M_{SD} d\sigma/dx ,\qquad (4.2)$$

where $\Delta\sigma$ is the shift in cm^{-1} due to Δx_{obj} (nm) and the linear dispersion $d\sigma/dx$ given by

$$\frac{d\sigma}{dx} = \frac{\cos(\beta)}{kn L_B \lambda^2}\qquad (4.3)$$

where λ is the wavelength, k the grating order, n the groove density, L_B the exit focal length and β the exit angle. Our experiments have shown that the measured frequency shifts are in agreement with the calculated $d\sigma/dx$ [1].

The effect of the focal spot size can not only be observed in the frequency but also in the intensity which is explained in the following chapter.

4.3 Size Dependent Spectral Line Width

Figure 4.3b shows the intensity of the G+ Raman mode of an individual single wall carbon nanotube as a function of exact position within the focal spot using two different objectives. The intensity is highest when the tube is on the optical axis and vanishes at the edge of the focal spot. This shows that the size of the focal spot can be probed by scanning a single nanotube in the spectral direction across the focal spot. A higher magnification of the optical objective results, as expected, in a smaller focal spot. The measured points agree with model calculations (shown by solid lines) in Fig. 4.3b.

This means the true wavelength of the emitter can only be determined by scanning the tube across the focal point in case its orientation with respect to the spectral direction is not known. This information can in turn be used to locate an individual carbon nanotube within a focal spot at a resolution greater than $\lambda/2$. When considering the $100\times/0.9$NA objective, the spectral position is determined to a precision of higher than a quarter of a wavenumber. Using the measured linear dispersion of $3.1\,\mathrm{cm}^{-1}/\mu\mathrm{m}$, the position of the nanotube can be determined to 80 nm-precision, or $\sim \lambda/12$. Assuming a second emitter with a non-overlapping spectral signature, such as a carbon nanotube of a different diameter, the distance between the two emitters in the spectral direction could then be determined with similar sub-wavelength precision.

Higher magnifications M_{OS} and M_{SD} increase the sensitivity to position and the linear dispersion is inversely proportional to the number of grooves, which means that a more dispersive grating will decrease the linear dispersion and decrease the sensitivity. However, since the pixel size of the array detector is constant, a higher dispersion can provide more data points of the spectral feature and thus allows the spectral position to be determined with greater precision.

In conventional spectroscopy, spectral resolution depends on size of the spectrometer entrance slit. The larger the slit the lower the resolution or in other words, a broader spectral line is obtained. Figure 4.4 shows the frequency width at half maximum as a function of the slit size for an extended emitter and an individual carbon nanotube. For the extended emitter, the line broadening increases linearly with slit size but this is not the case for the individual nanotube. For the nanotube emitter, the line broadening is constant and is independent of the slit size and is smaller than the line width of the extended emitter. This means that the size of the object has a direct effect on spectral line broadening or, differently put, the nano emitter itself acts as a nanometer sized slit. The nanotube having a diameter smaller than the focal spot is not influenced by the size of the spectrometer entrance slit which is in complete contrast to spectroscopy from an extended emitter.

The measured spectral line width from a macroscopic object is a convolution of the intrinsic line width with the instrument line width; the latter is the sum of the contributions from the width of the entrance slit and the detector pixel size. For a nano emitter, the image size at the slit is determined by the point spread function of the microscope objective. This image is mapped onto the detector and it is the

Fig. 4.4 Spectral width
(FWHM) of a single CNT
and an extended emitter (Ne
703.2 nm line) as a function
of the spectrometer entrance
slit). Reprinted with
permission from reference
[1], Copyright (2008)
American Chemical Society

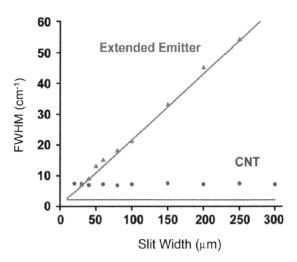

diffraction limited point spread function that determines the line width. The line width is independent of the slit width once the slit is larger than the point spread function and a fully open slit does not contribute to a broadening of the line shape for a nano emitter, in contrast to the case of a macroscopic object, where the slit width always contributes to the line width. The measured line width for the CNT emitter (when using the 50×/0.5NA objective) is shown in Fig. 4.4 by blue circles. The solid blue line shows the contribution due to limitation in the resolution of the instrument where the line width is approximated by the Rayleigh criterion. Subtracting the instrument contribution to the measured line width of 7.5 cm^{-1}, the intrinsic CNT Raman line width for the G band of a suspended nanotube is 6.9 cm^{-1}. We consider a case where the objective is focused on a 100 μm pinhole that is illuminated from below by a Ne lamp. The calculated contribution to line width for this system is shown by the solid green line.

We now describe what happens to the spectral signal when the nanoparticles are smaller than the size of the focal spot. Apparent spectral shifts as large as ± 5cm^{-1} can appear depending on the numerical aperture of the objective. To find the true spectral frequency of the nano emitter, spectral scans need to be carried out across the focal point in the spectral direction. The true spectral frequency then corresponds to the spectral position of maximum intensity. The exact location of the nano emitters can thus be determined with an accuracy of λ/12 using apparent spectral shifts. It can be said that the nano emitter acts as its own 'slit' and the spectral line broadening is limited by its intrinsic line width and the spectrometer entrance slit can be left open without increasing instrument line broadening.

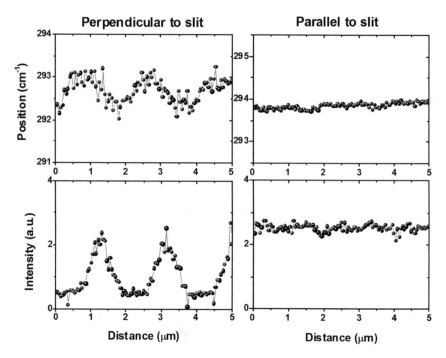

Fig. 4.5 Left side shows frequency and intensity of the spectral band when displacing the trenches in the spectral direction. The graph on the right shows the frequency and intensity when displacing in a direction perpendicular to the spectral direction. Reprinted with permission from reference [3], Copyright (2012) American Institute of Physics

4.4   Topography Induced Apparent Spectral Shifts

We have seen in the previous section that apparent spectral shifts occur because the emitter is smaller than the size of the focal spot. This implies that such apparent spectral shifts can appear for any emitter which is non-uniform at the scale of the focal spot size. Non-uniformity can also be caused by changes in topography where a part of the emitter could be in focus and another part out of focus. Below, we illustrate topography induced spectral shifts on a semiconductor surface with trenches created using ion etching or electron beam lithography.

Figure 4.5 shows Raman spectra of a gallium arsenide (GaAs) substrate with parallel grooves generated by electron beam lithography, where the edge widths are smaller than the focal spot size (<100 nm) [3]. We recorded Raman spectra of the longitudinal optical (LO) phonon of GaAs (symmetry E) while scanning the sample in a direction perpendicular to the direction of the trenches. Scans were performed both in the spectral direction and perpendicular to it. For both configurations, the incident light polarization was set parallel to the spectrometer grating direction. The sample was scanned in steps of 50 nm using a piezoelectric displacement table and an objective with a numerical aperture of 0.9.

When scanning in the spectral direction, the spectral peak position shifts on average by $0.53 \, \text{cm}^{-1}$ and when scanning perpendicular to the spectral direction we see that topography induced shifts are consistently smaller and oscillate on average by $0.21 \, \text{cm}^{-1}$, which is close to noise level. The same spectral shifts are observed when rotating the angle of polarization of the incident beam.

The spectral shift due to one side illumination can be estimated by multiplying the corresponding displacement with respect to the optical axis by the magnification of the optical spot at the entrance slit and the dispersion of the grating ($0.051 \, \text{cm}^{-1}/\mu\text{m}$) [3]. The observed shift of $0.53 \, \text{cm}^{-1}$ corresponds to a maximum displacement of 0.96 μm from the optical axis.

We find that the observed spectroscopic shift in Fig. 4.4 is smaller than what has been observed for individual carbon nanotubes. While for a nanoscale object the size is always the same and only its position with respect to the optical axis changes, here, the fraction illuminating the focal point depends on the position of the edge with respect to the optical axis, which ranges from fully illuminating the focal spot to only illuminating at the edge.

Topography changes imply that the surface is either higher or lower with respect to the focal spot and this obviously influences the intensity of the detected signal. Figure 4.6 shows the effect of over or under focusing of the objective lens on the recorded spectra. When the focal spot is too high, the oscillations of the spectral position are larger ($0.95 \, \text{cm}^{-1}$) compared to when the focal point is on the surface in the spectral direction. When the focal spot is lower, the oscillations of the spectral position are barely visible. This shows that the exact spectral position is also influenced by the vertical position of the focal spot with respect to the surface, see Fig. 4.6.

Figure 4.7 shows the elliptically shaped focal spot in a vertical plane. The scattering volume from which the signal originates in the left and right part of the focal spot is different when the center of the focal spot is above, on or below the surface. This influences the illumination of the two halves of the focal spot and hence the apparent spectral shift appears.

Topography induced spectral shifts can be studied in much greater detail by examining frequency shifts of the elastically scattered light (Rayleigh scattering). The signal is several orders of magnitude larger thus increasing the signal to noise ratio. To do this, we etched trenches that were 580 nm wide and 850 nm deep in two perpendicular directions in silicon using ion beams. The elastic spectral line was recorded with a microscope spectrometer system equipped with a 100×0.9 NA objective, krypton ion laser (530.9 nm) and 2400 lines/mm grating with a focal length of 640 mm (Horiba T64000 triple spectrometer). The elastic spectral line was recorded across 4–6 pixels on the detector, at a rate of 10–20 k counts/s corresponding to less than 1 μW of laser power [4].

Figure 4.8 shows the position of the focal spot with respect to the groove in the form of an ellipsoid. Away from the edges (positions 1, 3, 5) light within the focal spot is scattered symmetrically. There is no difference between the left and right half of the focal spot and no spectral shifts are observed. However, at the edges (positions

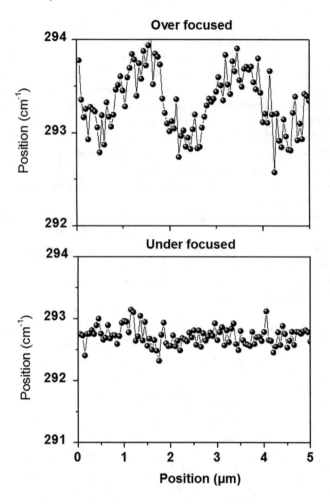

Fig. 4.6 Spectral shifts when displacing the trenches in the spectral direction under condition of over or under focus.Reprinted with permission from reference [3], Copyright (2012) American Institute of Physics

Fig. 4.7 Position of the focal spot (ellipsoid) with respect to the groove when in focus (I), over (O) and under focused (U)

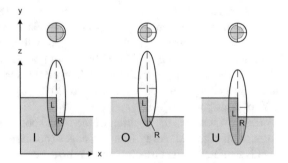

Fig. 4.8 Schematic of focal spot (FP) indicated as an ellipsoid with respect to the groove topography in the spectral direction. The focal point in positions 1,3 and 5 give rise to no frequency shift because the light scattered within the focal spot is distributed symmetrically. The focal spot in position 2 leads to positive spectral shifts while that in position 4 leads to negative frequency shifts because light scattered within the focal spot is distributed asymmetrically in one direction or the other

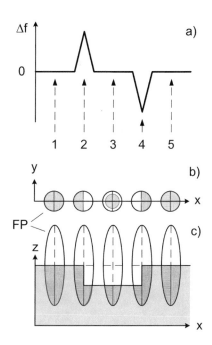

2, 4) light scattered within the focal spot is distributed asymmetrically and a spectral shift occurs due to geometrical reasons as pointed out earlier. The frequency shift is positive when the step is falling and negative when the step is rising.

Figure 4.9 shows the intensity variations (open circles) and spectral positions (filled circles) of the fitted Rayleigh peak when scanning in the spectral direction. The Rayleigh peak up shifts as the focal spot approaches the falling groove edge and down shifts as the focal spot approaches the rising edge of the groove. An additional peak is observed in between the edges when the beam is scanned through the bottom of the groove indicating a small falling edge at the bottom of the groove. The intensity is reduced at the groove bottom because the focal point is out of focus.

The magnitude of the apparent spectral shift can be determined using a simple model. The contribution to the spectral shift from each point in the focal plane is proportional to its distance from the optical axis, the dispersion of the grating, magnification, and light intensity. The maximum spectral shift $\sigma_{max} = 0.37$ cm^{-1} is due to an edge illuminated by half of the focal spot assuming Gaussian intensity distribution. This is consistent with a magnification between sample and detector of $M = 100$ and a grating dispersion, $d\sigma/dx = 0.0177$cm^{-1}/μm.

The focus is not changed during the scan and is initially on the surfaces outside the groove area. The lines in the figure are Gaussian fits to the measured intensity. The spectra at the two edges are clearly blue- and red-shifted compared to those outside and at the bottom of the groove and can be easily distinguished due to the high signal to noise ratio. We note that the edge width (130 nm) is smaller than the point spread function while the groove width and depth are larger than the wavelength

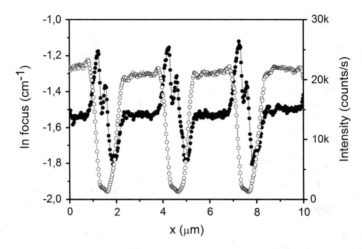

Fig. 4.9 Frequency (full circles, left axis) and intensity variations (empty circles, right axis) of the Rayleigh peak when scanning in the spectral direction across the grooves. Frequency variations are large at the groove edges. The peak in between corresponds to a small step at the groove bottom [4]

of the incident laser beam. When placing the focal spot on the groove edge, only one half of the focal spot is illuminated with the groove bottom being out of focus. The illuminated side of the focal spot forms an image on the detector plane, where again, only part of the image of the focal spot is illuminated. Thus, asymmetric or off-axis illumination in the spectral direction results in a change in the grating angle and leads to a spectral red or blue shift depending on whether the left or right half of the focal spot is illuminated.

From Fig. 4.9 it can be seen that the spectral shift is close to the derivative of the intensity function (Fig. 4.9 open circles) when scanning the samples through the focal spot in the spectral direction. The maximum and minimum of the spectral shift are separated by a lateral displacement of 633 nm (± 8 nm), which is close to the width of the groove (582 nm) as measured from SEM images. The FWHM of the intensity function is 900 nm. Hence, a 30% increase in edge resolution can be obtained by exploiting the variation of the apparent spectral shift.

Thus, variations at scales smaller than the size of the focal spot have several consequences for optical spectroscopy of nano emitters. Apparent spectral shifts which can be as high as several wave numbers occur in the spectral direction. For small nanoparticles, the spectral line width is reduced and depends on the size of the emitter. The position of a nanoparticle such as single wall carbon nanotubes can be determined with an accuracy of $\lambda/12$. Over or under focusing influences the apparent spectral shift. In the case of topography variations at scales smaller than the size of the focal spot, the apparent spectral shift is proportional to the first derivative of the intensity function. This permits improvement in the determination of the step edge by around 30%.

The presence of apparent spectral shifts points to the fact that spectrometer calibration depends on the roughness of the substrate and could be a source of error in the determination of spectral position. It is also noted that optical spectra recorded from nanoparticles contain intrinsic errors due to their size being smaller than that of the focal spot. To find the true spectral position the nanoparticle needs to be scanned across the focal spot in the spectral direction.

References

1. A.G. Walsh, W.S. Bacsa, A.N. Vamivakas, A.K. Swan, Spectroscopic properties unique to nano-emitters. Nano Lett. **8**, 4330 (2008)
2. J.E. Colwell, B.M. Jakosky, Effects of topography on thermal infrared spectra of planetary Surfaces. J. Geophys. Res. **107**(11), 5106 (2002)
3. V. Tishkova, W.S. Bacsa, Apparent Raman spectral shifts from nano-structured surfaces. Appl. Phys. Lett. **100**, 173105 (2012)
4. E. Pavlenko, P. Salles, R.R. Bacsa, W.S. Bacsa, Topography induced apparent spectral shifts and implications for optical nano-spectroscopy, submitted for publication (2020)

Chapter 5
Microscopic Origin of the Index of Refraction

5.1 Introduction

We saw in Chap. 1 that the interaction of light with matter can be described by light
scattering on each atom. Since the wavelength of light is much larger than the size of
the atom, the atom can be described as an electron oscillator driven by the incident
field. The wave scattered by the atom depends on the resonant frequency of the
oscillator which in turn determines the phase and amplitude of the scattered wave.
The incident wave polarizes the atom giving rise to a dipole wave that propagates in
all directions except in the direction of polarization. In the case of an ensemble of
atoms such as in matter, many scattering events occur. Since the electromagnetic wave
propagates at a finite speed, scatterers separated in space emit waves which are phase
shifted with respect to each other. The superposition of all the scattered waves with
the incident wave thus gives rise to an accumulated phase change. This accumulated
phase change leads to a reduction of the speed of light in matter. The reflected and
transmitted beams result from constructive interference of all the scattered waves
with the incident beam.

Here our goal is to connect our macroscopic understanding of the phenomenon of
reflection and refraction to what happens at the atomic level by integrating the emitted
waves from all the atoms. Since the size of the atom is 3 orders of magnitude smaller
than the wavelength of light, we can safely assume the presence of a continuum of
dipole waves. When integrating over all the scattered waves it is important to take
into account the phase shift due to the finite propagation speed of light. That means
two points in the material will emit scattered waves at two different times which is
taken into account by the retarded time.

Both Schwartz [2] and Feynman [3] considered the interaction of radiation with
matter using an incident plane wave perpendicular to the surface. Under certain
assumptions they were able to calculate the field inside the medium. More recently
Lai et al. [1] used an oblique angle of incidence which circumvents the difficulties
encountered when considering a normal angle of incidence. Here we follow the
approach of Lai et al. and give more details to clarify the calculation.

W. Bacsa et al., *Optics Near Surfaces and at the Nanometer Scale*,
SpringerBriefs in Physics, https://doi.org/10.1007/978-3-030-58983-7_5

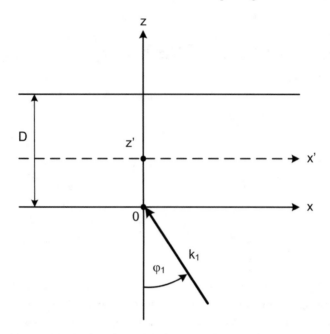

Fig. 5.1 Incident wave and a slab of material of thickness D. A thin sheet is located at z' parallel to the two interfaces

5.2   Electric Field Generated by Illuminating a Sheet of Material

To understand the electric field generated by a material we consider a slab of thickness D which is illuminated from the lower surface by a light wave. The material may then be considered to occupy $z \in [0, D]$, as shown in Fig. 5.1. The incident field causes a current in the ensemble of atoms. This current generates a scalar and vector potential from which the electric and magnetic fields of incident, refracted and reflected waves can be calculated.

When expressed in terms of the scalar and vector potentials the four Maxwell's equations can be reduced to two

$$\frac{\partial^2 \phi}{\partial t^2} - c^2 \left(\frac{\partial^2 \phi}{\partial x^2} + \frac{\partial^2 \phi}{\partial y^2} + \frac{\partial^2 \phi}{\partial z^2} \right) = \frac{\rho}{\epsilon_0} \tag{5.1}$$

$$\frac{\partial^2 \mathbf{A}}{\partial t^2} - c^2 \left(\frac{\partial^2 \mathbf{A}}{\partial x^2} + \frac{\partial^2 \mathbf{A}}{\partial y^2} + \frac{\partial^2 \mathbf{A}}{\partial z^2} \right) = \mu_0 \mathbf{J} \tag{5.2}$$

where $\mathbf{A}, \phi$ are the magnetic and electric potentials, ρ is the charge density and $\mathbf{J}$ the current density. The potentials are related to the electric and magnetic field by the relations

$$\mathbf{E} = -\nabla \phi - \frac{\partial \mathbf{A}}{\partial t}, \qquad \mathbf{B} = \nabla \times \mathbf{A}. \tag{5.3}$$

For the present problem we consider a single-frequency electromagnetic wave acting on a linear medium. The field interacts with the charges in the medium, such as the electrons, which then generate further potentials. Since it takes a finite time for the resultant waves to travel any distance we introduce the concept of a *retarded time* t_r. We define this by considering an observer at point $\mathbf{r}$. They observe a wave at time t which was originally emitted from a point $\mathbf{r}'$. The time taken to travel this distance is $|\mathbf{r} - \mathbf{r}'|/c$. Since the wave reaches them at time t the time when the wave was actually released, the retarded time, is therefore $t_r = t - |\mathbf{r} - \mathbf{r}'|/c$.

The appropriate solution to the above potential equations, on a domain V, may be expressed in terms of the retarded time t_r,

$$\mathbf{A}(\mathbf{r}, t) = \frac{\mu_0}{4\pi} \int_V \frac{\mathbf{J}(\mathbf{r}', t_r)}{|\mathbf{r} - \mathbf{r}'|} \, dV' \tag{5.4}$$

$$\phi(\mathbf{r}, t) = \frac{1}{4\pi \epsilon_0} \int_V \frac{\rho(\mathbf{r}', t_r)}{|\mathbf{r} - \mathbf{r}'|} \, dV'. \tag{5.5}$$

These may be viewed as the mathematical statement that the potential at a given point is due to the sum of densities throughout the domain. The influence of each density diminishes with the inverse of the distance. Since the material is homogeneous and not charged, $\rho = 0$, we may neglect ϕ within the medium. However at the boundaries there is an inhomogeneity, where the material ends and the new one begins, there we will have to deal with surface charges. So, in the following analysis when dealing with the Transverse Electric wave we will neglect the charge density but in the case of the Transverse Magnetic wave we will include a contribution at the outer surfaces of the material.

If the incident wave has a time-dependence of the form $e^{i\omega t}$ then we assume that the charges in the medium also emit waves with this time-dependence. We may then relate the time-dependence of the wave in terms of t or t_r via

$$\exp(i\omega t_r) = \exp(i\omega(t - |\mathbf{r} - \mathbf{r}'|/c)) = \exp(i\omega t)\exp(-ik|\mathbf{r} - \mathbf{r}'|) \tag{5.6}$$

where $k = \omega/c$. To shorten expressions in the following we will often separate the time and x dependent terms. Here we isolate the time dependence via

$$\mathbf{J}(\mathbf{r}', t_r) = \mathbf{J}(\mathbf{r}')\exp(i\omega t_r) = \mathbf{J}(\mathbf{r}')\exp(-ik|\mathbf{r} - \mathbf{r}'|)\exp(i\omega t), \tag{5.7}$$

(and similar for ρ) which allows us to write Eqs. (5.4) and (5.5) as

$$\mathbf{A}(\mathbf{r})\,\exp(i\omega t) = \frac{\mu_0}{4\pi}\int_{V'}\frac{\mathbf{J}(\mathbf{r}')}{|\mathbf{r}-\mathbf{r}'|}\exp(-ik|\mathbf{r}-\mathbf{r}'|)\,dV'\,\exp(i\omega t) \qquad (5.8)$$

$$\phi(\mathbf{r})\,\exp(i\omega t) = \frac{1}{4\pi\epsilon_0}\int_{V'}\frac{\rho(\mathbf{r}')}{|\mathbf{r}-\mathbf{r}'|}\exp(-ik|\mathbf{r}-\mathbf{r}'|)\,dV'\,\exp(i\omega t)\,, \qquad (5.9)$$

(recall $\rho(\mathbf{r}') = 0$ within the material).

Now we focus on the case where a plane wave interacts with a thin sheet of thickness $\delta z'$ located at z', as shown in Fig. 5.1. In the transverse electric (TE) state we may orient the x, y axes such that the wave vector is $\mathbf{k} = (k_x, 0, k_z)$, with the dispersion relation $k_x^2 + k_z^2 = k^2 = \omega^2/c^2$. The wave may be written as

$$\mathbf{E}_{inc} = \mathbf{E}_0 \exp\left(i\left(\omega t - k_x x - k_z z\right)\right)\,. \qquad (5.10)$$

Note, to conform with previous literature here we write $\mathbf{E} = \mathbf{E}_0 \exp(i(\omega t - \mathbf{k} \cdot \mathbf{r}))$ rather than $\mathbf{E} = \mathbf{E}_0 \exp(i(\mathbf{k} \cdot \mathbf{r} - \omega t))$ as used in previous chapters. However, since our interest lies in the physical solution, which corresponds to the real part, and cosine is an even function, $\cos(x) = \cos(-x)$, the results are identical irrespective of the chosen form. The wave vector has (x, z) components hence we take the field vector in the y direction, $\mathbf{E}_0 = E_0\hat{\mathbf{y}}$. In the corresponding Transverse Magnetic (TM) case the field also has x and z components. When the wave interacts with the sheet, which is located at a fixed z plane, the tangential component of the wave vector is unaffected however the normal component changes (hence we observe refraction). Since the incoming wave takes the form of Eq. (5.10) the resultant wave produced by the interaction between the incoming wave and material will be

$$\mathbf{E} = \exp\left(i\left(\omega t - k_x x\right)\right)\left(E_1 \exp(ik_{mz}z) + E_2 \exp(-ik_{mz}z)\right)\hat{\mathbf{y}}\,, \qquad (5.11)$$

where E_1, E_2, k_{mz} are unknown. Equation (5.10) has a $e^{-ik_z z}$ dependence since the wave is travelling upwards. In (5.11) we have $e^{\pm ik_{mz}z}$ since the resultant wave will be emitted in both the positive and negative z direction. Note, while here we anticipate the form of the resultant wave we will subsequently find it appears naturally through the governing equations.

In the following we will replace the current density using the standard relation between current density and electric field. The common approximation is $\mathbf{J} = \sigma\mathbf{E}$ where σ is the electrical conductivity. With free currents this form of Ohm's law has σ as the electrical conductivity. However, most materials have free as well polarization charges so we should also consider the contribution due to the movement of electric dipole moments

$$\mathbf{J}_p = \frac{\partial\mathbf{P}}{\partial t} \qquad (5.12)$$

where $\mathbf{P} = \epsilon_0\chi_e\mathbf{E}$ is the polarization and χ_e is the *electric susceptibility* (a measure of how much the material is polarized by an electric field). In a magnetic material we also find magnetic currents

$$\mathbf{J}_M = \nabla \times \mathbf{M} . \tag{5.13}$$

Combined, these two terms form the *bound current*. The total current is then the sum of free and bound currents. In the following we assume the material is non-magnetic and so the total current is

$$\mathbf{J} = \sigma \mathbf{E} + \epsilon_0 \chi_e \frac{\partial \mathbf{E}}{\partial t} = (\sigma + i\omega\epsilon_0 \chi_e) \mathbf{E} = \sigma_f \mathbf{E} , \tag{5.14}$$

where the *effective electrical conductivity* $\sigma_f = \sigma + i\omega\epsilon_0 \chi_e$. In a similar manner we may define an effective susceptibility

$$\mathbf{J} = i\omega\epsilon_0 \chi_f \mathbf{E} , \tag{5.15}$$

where $\chi_f = \chi_e - i\sigma/(\omega\epsilon_0)$. The two relations indicate that the material can be considered as either conducting or insulating. The current density in a sheet subject to the incoming wave (5.10) will then be

$$\mathbf{J}(\mathbf{r}, t') = \sigma_f \exp\left(i\left(\omega t - k_x x\right)\right) \left(E_1 \exp(ik_{mz}z) + E_2 \exp(-ik_{mz}z)\right) \hat{\mathbf{y}} . \tag{5.16}$$

From now on, to reduce the length of the mathematical expressions, we deal with the time-independent quantity

$$J(\mathbf{r}) = \exp\left(-ik_x x\right) J_z(z) , \tag{5.17}$$

where

$$J_z(z) = \sigma_f \left(E_1 \exp(ik_{mz}z) + E_2 \exp(-ik_{mz}z)\right) \tag{5.18}$$

comprises the part of J which varies with z. Figure 5.2 shows how the incident wave modulates the electric field in x direction.

Away from the thin sheet we choose a random point (x, y, z) and then, for mathematical simplicity, take the (x', y') origin on the sheet as being directly above or below this (depending on whether $z' > z$ or $< z$). Since the electric potential is zero inside the medium the electric field induced at (x, y, z) by the whole sheet may be calculated from Eq. (5.3) with $\phi = 0$, that is $\mathbf{E} = -\partial \mathbf{A}/\partial t$. So we now focus on the vector potential.

From Eq. (5.8) we see that the vector potential inside the thin sheet is defined by

$$\delta \mathbf{A}(\mathbf{r}) = \frac{\mu_0}{4\pi} \hat{\mathbf{y}} \int_{y'} \int_{x'} \frac{1}{|\mathbf{r} - \mathbf{r}'|} e^{-ik_x x'} e^{-ik|\mathbf{r}-\mathbf{r}'|} dx' dy' \left(J_z(z')\delta z'\right) \tag{5.19}$$

where

$$|\mathbf{r} - \mathbf{r}'| = \sqrt{(x - x')^2 + (y - y')^2 + (z - z')^2} . \tag{5.20}$$

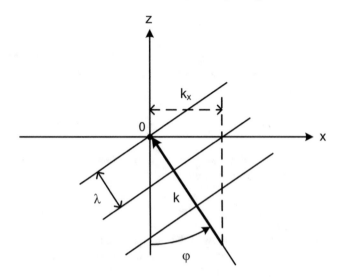

Fig. 5.2 At non-normal incidence, the electric field oscillates not only in time but also along x in a plane perpendicular to z. The oscillations along x are given by k_x

We now focus on the integral

$$I = \int_{y'} \int_{x'} \frac{1}{|\mathbf{r} - \mathbf{r}'|} \exp(-ik_x x') \exp(-ik|\mathbf{r} - \mathbf{r}'|) \, dx' dy' . \tag{5.21}$$

This is more easily tackled by shifting to a polar co-ordinate system, as shown in Fig. 5.3. Any point (x', y') located within the sheet may be expressed as $(x + \rho' \cos \theta', y + \rho' \sin \theta')$, i.e. the radial system has origin at (x, y, z'). The integration area and distance may be expressed as

$$dx' dy' = \rho' d\rho' d\theta' \tag{5.22}$$

$$R' = |\mathbf{r} - \mathbf{r}'| = \sqrt{\rho'^2 + (z - z')^2} . \tag{5.23}$$

Further, since both the position of the sheet and the point (x, y, z) are fixed the relation $R'^2 = \rho'^2 + (z - z')^2$ indicates $R' dR' = \rho' d\rho'$. If $\rho' \in [0, \infty]$ then $R' \in [|z - z'|, \infty]$ and

$$I = \int_0^\infty \int_0^{2\pi} \frac{1}{R'} \exp(-ik_x(x + \rho' \cos \theta')) \exp(-ikR') \, \rho' d\theta' d\rho' \tag{5.24}$$

$$= \exp(-ik_x x) \int_{|z-z'|}^\infty \int_0^{2\pi} \exp(-ik_x \sqrt{R'^2 - (z - z')^2} \cos \theta') \exp(-ikR') \, d\theta' dR'.$$

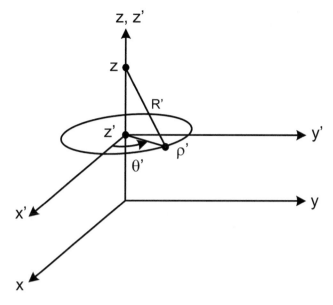

Fig. 5.3 Points in the sheet of material in the plane (x', y') are described by θ' and ρ'. R' is the distance between the observation point on axis z and a point in the sheet

Before proceeding we need a short digression on some established integral relations.

Poisson's Integral for Bessel Functions

Poisson's integral representation for the Bessel functions J_ν states

$$J_\nu(\alpha) = \frac{1}{\Gamma(\nu + 1/2)\sqrt{\pi}} \left(\frac{\alpha}{2}\right)^\nu \int_0^\pi \exp(i\alpha \cos s)(\sin s)^{2\nu} \, ds$$

$$= \frac{1}{\Gamma(\nu + 1/2)\sqrt{\pi}} \left(\frac{\alpha}{2}\right)^\nu \int_0^\pi \cos(\alpha \cos s)(\sin s)^{2\nu} \, ds \quad \Re(\nu) > -1/2 \,,$$

where the Γ function is the extension of the factorial function to real numbers. For integer n, $\Gamma(n) = (n-1)!$ Noting that $\Gamma(1/2) = \sqrt{\pi}$ we obtain an integral representation for the zeroth order Bessel function

$$J_0(\alpha) = \frac{1}{\pi} \int_0^\pi \exp(i\alpha \cos s) \, ds = \frac{1}{\pi} \int_0^\pi \cos(\alpha \cos s) \, ds \,.$$

Our interest lies with an integral between $0, 2\pi$. Defining $s' = s + \pi$ gives

$$\pi J_0(\alpha) = \int_\pi^{2\pi} \exp(i\alpha \cos(s' - \pi)) \, ds' = \int_\pi^{2\pi} \exp(-i\alpha \cos s') \, ds' .$$

For even n the Bessel J functions are even and so we may change the sign of α in the integral

$$\pi J_0(\alpha) = \pi J_0(-\alpha) = \int_\pi^{2\pi} \exp(-i\alpha \cos s') \, ds' = \int_\pi^{2\pi} \exp(i\alpha \cos s') \, ds' .$$

Adding integrals over $(0, \pi)$ and $(\pi, 2\pi)$ and changing the dummy variable to s in the second, we may write

$$2\pi J_0(\alpha) = \int_0^\pi \exp(i\alpha \cos s) \, ds + \int_\pi^{2\pi} \exp(i\alpha \cos s) \, ds$$

$$= \int_0^{2\pi} \exp(i\alpha \cos s) \, ds . \tag{5.25}$$

Integrals of Combined Bessel and Trigonometric Functions

The classic text [4] provides a comprehensive list of standard integrals. To evaluate I we need the following two relations

$$\int_a^\infty J_0(b\sqrt{s^2 - a^2}) \sin(cs) \, ds = \frac{\cos(a\sqrt{c^2 - b^2})}{\sqrt{c^2 - b^2}} \qquad 0 < b < c \tag{5.26}$$

$$\int_a^\infty J_0(b\sqrt{s^2 - a^2}) \cos(cs) \, ds = -\frac{\sin(a\sqrt{c^2 - b^2})}{\sqrt{c^2 - b^2}} \qquad 0 < b < c , \tag{5.27}$$

see [4, Eqs. 6.677 (1), (2)].

Taking the result (5.25) with $\alpha = -k_x\sqrt{R'^2 - (z - z')^2}$ and noting that J_0 is even, we may evaluate the θ' integral of I to find

$$I = \exp(-ik_x x) \int_{|z-z'|}^\infty \exp(-ikR') \, 2\pi J_0(k_x\sqrt{R'^2 - (z - z')^2}) \, dR' . \tag{5.28}$$

Splitting the R' exponential into sine and cosine terms and then setting $a = |z - z'|$, $b = k_x, c = k$ (since $k_x < k$ then $b < c$ is satisfied) in the relations (5.26), (5.27) we obtain

$$\int_{|z-z'|}^{\infty} J_0(k_x\sqrt{R'^2-(z-z')^2})e^{-ikR'}\,dR' = -\frac{\sin(|z-z'|\sqrt{k^2-k_x^2})}{\sqrt{k^2-k_x^2}}$$

$$-i\frac{\cos(|z-z'|\sqrt{k^2-k_x^2})}{\sqrt{k^2-k_x^2}} = -\frac{i}{k_z}e^{-ik_z|z-z'|}\,, \tag{5.29}$$

after applying $k^2 - k_x^2 = k_z^2$. Finally we obtain the remarkably simple formula

$$I = -\frac{2\pi i}{k_z}e^{-i(k_xx+k_z|z-z'|)}\,. \tag{5.30}$$

Replacing the integral back into Eq. (5.19) we find an expression for the vector potential at a point (x, y, z) induced by a sheet of thickness $\delta z'$ at position $z' < z$ (i.e. $|z - z'| = z - z'$)

$$\delta\mathbf{A}^+(\mathbf{r}) = \frac{\mu_0}{4\pi}\hat{\mathbf{y}}IJ_z(z')\delta z' = -\frac{\mu_0 i}{2k_z}\hat{\mathbf{y}}e^{-i(k_xx+k_z(z-z'))}J_z(z')\delta z'\,. \tag{5.31}$$

The induced electric field when $z > z'$ is therefore

$$\delta\mathbf{E}_A^+(\mathbf{r}, t) = -\frac{\partial}{\partial t}(\delta\mathbf{A}^+(\mathbf{r}, t)) = -i\omega\delta\mathbf{A}^+(\mathbf{r})e^{i\omega t} \tag{5.32}$$

and so

$$\delta\mathbf{E}_A^+(\mathbf{r}) = -\frac{\mu_0\omega}{2k_z}\hat{\mathbf{y}}e^{-i(k_xx+k_zz)}e^{ik_zz'}J_z(z')\,\delta z'\,. \tag{5.33}$$

The corresponding potential and field when $z < z'$ follow in the same manner (where now $|z - z'| = -(z - z')$)

$$\delta\mathbf{A}^-(\mathbf{r}) = -\frac{\mu_0 i}{2k_z}\hat{\mathbf{y}}e^{-i(k_xx-k_z(z-z'))}J_z(z')\,\delta z' \tag{5.34}$$

$$\delta\mathbf{E}_A^-(\mathbf{r}) = -\frac{\mu_0\omega}{2k_z}\hat{\mathbf{y}}e^{-i(k_xx-k_zz)}e^{-ik_zz'}J_z(z')\,\delta z'\,. \tag{5.35}$$

Now assume we have a continuous medium divided into many of the sheets discussed above. The total electric field is the sum of the incident and all the induced fields for sheets above ($z' > z$) and below ($z' < z$) the reference point

$$\mathbf{E} = \mathbf{E}_{inc} + \Sigma\delta\mathbf{E}_A^- + \Sigma\delta\mathbf{E}_A^+\,. \tag{5.36}$$

Allowing the sheet thickness to tend to zero and then summing over infinitely many we arrive at the integral formulation. This will be discussed below for the Transverse Electric and Transverse Magnetic polarization.

5.3   Transverse Electric (TE) Mode

The total electric field is the sum of the incoming wave, all slices with $z > z'$ and all slices with $z < z'$. For the TE wave, if the slice is below z, that is $z' \in [0, z]$ then the contribution comes only from Eq. (5.33). If the slice is above z, so that $z' \in [z, D]$, then the contribution is from Eq. (5.35). Hence the total field

$$\mathbf{E}(\mathbf{r}) = \mathbf{E}_{inc} + \int_0^z \delta\mathbf{E}^+ \, dz' + \int_z^D \delta\mathbf{E}^- \, dz' \tag{5.37}$$

$$= \mathbf{E}_{inc} - \frac{\mu_0\omega}{2k_z} e^{-ik_x x} \hat{\mathbf{y}} \left(e^{-ik_z z} \int_0^z e^{ik_z z'} J_z(z') \, dz' + e^{ik_z z} \int_z^D e^{-ik_z z'} J_z(z') \, dz' \right) .$$

The function $J_z(z')$, given by Eq. (5.18), is simply the sum of two exponentials so it is a simple matter to evaluate the integrals, however the resultant expressions are rather long. A useful first approach is instead to examine the ordinary differential equation form (see [2]). For this we need to use Leibniz' theorem, which provides the rules for differentiating through an integral.

Leibniz' Theorem

$$\frac{d}{dz} \int_{a(z)}^{b(z)} f(z, z') \, dz' = f(b)\frac{db}{dz} - f(a)\frac{da}{dz} + \int_a^b \frac{\partial}{\partial z} f(z, z') \, dz'. \tag{5.38}$$

Ohm's law allows us to write $E_z(z) = \sigma_f J_z(z)$, then the z dependent component of (5.37) may be written

$$E_z(z) = E_0 e^{-ik_z z} \tag{5.39}$$

$$- \frac{\mu_0\omega\sigma_f}{2k_z} \left(e^{-ik_z z} \int_0^z e^{ik_z z'} E_z(z') \, dz' + e^{ik_z z} \int_z^D e^{-ik_z z'} E_z(z') \, dz' \right) .$$

Using Leibniz' theorem we may reduce this integral equation to an ordinary differential equation which is easily solved. We begin by differentiating equation (5.39)

$$\frac{d}{dz} \left(e^{-ik_z z} \int_0^z e^{ik_z z'} E_z(z') \, dz' \right)$$

$$= -ik_z e^{-ik_z z} \int_0^z e^{ik_z z'} E_z(z') \, dz' + e^{-ik_z z} e^{ik_z z} E_z(z)$$

$$= -ik_z e^{-ik_z z} \int_0^z e^{ik_z z'} E_z(z') \, dz' + E_z(z) \tag{5.40}$$

and then

$$\frac{d^2}{dz^2}\left(e^{-ik_zz}\int_0^z e^{ik_zz'}E_z(z')\,dz'\right)$$
$$= -k_z^2 e^{-ik_zz}\int_0^z e^{ik_zz'}E_z(z')\,dz' - ik_z E_z(z) + \frac{dE_z}{dz}. \tag{5.41}$$

Similarly

$$\frac{d^2}{dz^2}\left(e^{ik_zz}\int_0^z e^{-ik_zz'}E_z(z')\,dz'\right)$$
$$= -k_z^2 e^{ik_zz}\int_z^D e^{-ik_zz'}E_z(z')\,dz' - ik_z E_z(z) - \frac{dE_z}{dz}. \tag{5.42}$$

The second derivative of Eq. (5.39) is then

$$\frac{d^2 E_z}{dz^2} = -k_z^2 E_0 e^{-ik_zz} - \frac{\mu_0\omega\sigma_f}{2k_z}\left(-k_z^2\left\{e^{-ik_zz}\int_0^z e^{ik_zz'}E_z(z')\,dz'\right.\right.$$
$$\left.\left. +e^{ik_zz}\int_z^D e^{-ik_zz'}E_z(z')\,dz'\right\} - 2ik_z E_z(z)\right). \tag{5.43}$$

The first three terms on the right hand side coincide with the definition of $E_z(z)$, Eq. (5.39), multiplied by $-k_z^2$ and so we obtain

$$\frac{d^2 E_z}{dz^2} = -k_z^2 E_z - \frac{\mu_0\omega\sigma_f}{2k_z}(-2ik_z E_z) = -(k_z^2 - i\mu_0\omega\sigma_f)E_z. \tag{5.44}$$

The beauty of this format is that we may easily solve the differential equation to obtain

$$E_z(z) = E_1 e^{ik_{mz}z} + E_2 e^{-ik_{mz}z} \tag{5.45}$$

where E_1, E_2 are unknown constants and

$$k_{mz}^2 = k_z^2 - i\mu_0\omega\sigma_f. \tag{5.46}$$

To achieve this we did not make any assumption about the z dependence of J_z or E_z, however this is exactly the solution form anticipated by Eq. (5.11). Hence we have verified that assumption as well as identified the wavenumber k_{mz}.

Setting $k_{mz} = a - ib$, where a, b are real, we find

$$a = \pm\left(\frac{k_z^2 + \sqrt{k_z^4 + (\mu_0\omega\sigma_f)^2}}{2}\right)^{1/2}, \qquad b = \frac{\mu_0\omega\sigma_f}{2a}. \tag{5.47}$$

Since a is real we have neglected the $-\sqrt{k_z^4 + (\mu_0\omega\sigma_f)^2}$ option. Without loss of generality we may take a as positive, that is, the real part of k_m is positive (if we take the negative root then the results below simply swap between E_1 and E_2). So, within the medium the original wave, with wave number k_z in the z direction, now moves with a complex component k_{mz}. In the limit of a semi-infinite material $D \to \infty$, for the solution to decay requires $E_1 \to 0$: we will demonstrate this later. However in general there are two unknown quantities.

5.3.1 Index of Refraction

We are now in a position to determine the index of refraction, n, within the medium. The simplest approach is to use the definition in terms of the susceptibility

$$n^2 = 1 + \chi_f . \tag{5.48}$$

In Eq. (5.15) we defined the effective susceptibility, hence

$$n^2 = 1 + \chi_e - \frac{i\sigma}{\omega\epsilon_0} , \tag{5.49}$$

where σ, χ_e, are the electrical conductivity and susceptibility of the medium. Writing $n = n_r - in_i$ and then separating the real and imaginary parts of Eq. (5.49) we obtain

$$n_r = \pm \left(\frac{1 + \chi_e + \sqrt{(1 + \chi_e)^2 + \sigma^2/(\epsilon_0^2\omega^2)}}{2} \right)^{1/2} , \qquad n_i = \frac{\sigma}{2\epsilon_0\omega n_r} . \tag{5.50}$$

Alternatively we may use the definition

$$|\mathbf{k}_m|^2 = n^2 \frac{\omega^2}{c^2} \tag{5.51}$$

where

$$\begin{aligned} |\mathbf{k}_m|^2 = k_{mz}^2 + k_x^2 &= k_x^2 + k_z^2 - i\mu_0\omega\sigma_f \\ &= \frac{\omega^2}{c^2} - i\mu_0\omega(\sigma + i\omega\epsilon_0\chi_e) \\ &= \frac{\omega^2}{c^2}\left(1 + \chi_e - \frac{i\sigma}{\omega\epsilon_0}\right) \end{aligned} \tag{5.52}$$

and we have used the definition of σ_f from Eq. (5.14) and $\epsilon_0\mu_0 = 1/c^2$. As expected this coincides with the definition of (5.49).

If we note that the wave vector may be written in the form $|\mathbf{k}_m| = n\omega/c$ multiplied by a unit vector $\hat{\mathbf{u}}$, so $\mathbf{k}_m = (n\omega/c)\hat{\mathbf{u}}$, then the wave takes the form

$$\exp(-i\mathbf{k}_m \cdot \mathbf{r}) = \exp(-n_i(\omega/c)\hat{\mathbf{u}} \cdot \mathbf{r}) \, \exp(-in_r(\omega/c)\hat{\mathbf{u}} \cdot \mathbf{r}) \, . \qquad (5.53)$$

From this we see that the real part n_r is the refractive index that we would observe for the material. We define it via the first of Eqs. (5.50), taking the positive root (since the refractive index is positive for most standard materials). The complex part $n_i \geq 0$ shows how the wave decays as it passes through the medium. For this reason n_i is termed the extinction coefficient.

The complex refractive index is exactly the index that would be obtained by substituting a wave form $\exp(i(\omega t - \mathbf{k}_m \cdot \mathbf{r}))$ into Maxwell's equation. (Indeed we wrote down a less general form in Eq. (2.75), where we first discussed the decay of a wave inside a material). This would actually be a much simpler exercise but, in keeping with the theme of this monograph, we have now found it through the potential formulation accounting for the time taken for the wave to reach an observer. Using this approach we can see the microscopic origin of the observed refractive index. To verify this statement consider Eqs. (1.3) and (1.4). First we note

$$\nabla \times \nabla \times \mathbf{E} = \nabla(\nabla \cdot \mathbf{E}) - \nabla^2 \mathbf{E} = -\nabla^2 \mathbf{E} \qquad (5.54)$$

(recall, within the medium there is no charge, $\nabla \cdot \mathbf{E} = \rho/\epsilon_0 = 0$). Taking the curl of (1.3) we find

$$\nabla^2 \mathbf{E} = \frac{\partial}{\partial t}(\nabla \times \mathbf{B}) \, . \qquad (5.55)$$

Substituting for $\nabla \times \mathbf{B}$ from (1.4) leads to

$$\nabla^2 \mathbf{E} = \mu_0 \frac{\partial}{\partial t}\left(\mathbf{J} + \epsilon_0 \frac{\partial \mathbf{E}}{\partial t}\right) \, . \qquad (5.56)$$

With an electric field of the form (5.11) and $\mathbf{J} = \sigma_f \mathbf{E}$, Eq. (5.56) gives

$$\left(-k_x^2 - k_{mz}^2\right)\mathbf{E} = \left(\mu_0\sigma_f i\omega - \mu_0\epsilon_0\omega^2\right)\mathbf{E} \, . \qquad (5.57)$$

Noting that $\mu_0\epsilon_0 = 1/c^2$ and $k_x^2 + k_z^2 = \omega^2/c^2$ we recover Eq. (5.46) for the modified wavenumber $k_{mz}^2 = k_z^2 - i\mu_0\sigma_f\omega$.

5.3.2 The Electric Field Within the Medium

We already know the form of the electric field within the medium, this was written down way back in Eq. (5.11). However, that expression contains three unknowns,

E_1, E_2, k_{mz}. We have found an expression for k_{mz}, Eq. (5.46), so will now work on the other two coefficients. In fact we could have found them earlier by simply integrating equation (5.39) and comparing coefficients. So now we return to Eq. (5.39). Substituting for $E_z(z')$, via Eq. (5.45), leads to

$$E_z(z) = E_0 e^{-ik_z z} - \frac{\mu_0 \omega \sigma_f}{2k_z} \left[e^{-ik_z z} \int_0^z E_1 e^{i(k_{mz}+k_z)z'} + E_2 e^{-i(k_{mz}-k_z)z'} \, dz' \right.$$
$$\left. + e^{ik_z z} \int_z^D E_1 e^{i(k_{mz}-k_z)z'} + E_2 e^{-i(k_{mz}+k_z)z'} \, dz' \right]. \tag{5.58}$$

Evaluating the integrals

$$E_z(z) = E_1 e^{ik_{mz}z} + E_2 e^{-ik_{mz}z} = E_0 e^{-ik_z z} - \frac{\mu_0 \omega \sigma_f}{2k_z} \Lambda \tag{5.59}$$

where

$$\Lambda = \frac{E_1}{i(k_{mz}+k_z)} \left(e^{ik_{mz}z} - e^{-ik_z z} \right) - \frac{E_2}{i(k_{mz}-k_z)} \left(e^{-ik_{mz}z} - e^{-ik_z z} \right)$$
$$+ \frac{E_1}{i(k_{mz}-k_z)} \left(e^{i(k_{mz}-k_z)D} e^{ik_{mz}z} - e^{ik_{mz}z} \right)$$
$$- \frac{E_2}{i(k_{mz}+k_z)} \left(e^{-i(k_{mz}+k_z)D} e^{ik_z z} - e^{-ik_{mz}z} \right). \tag{5.60}$$

Noting that $k_{mz}^2 - k_z^2 = -i\omega\mu_0\sigma_f$ this may be rearranged to

$$\Lambda = -\frac{2k_z}{\omega\mu_0\sigma_f}(E_1 e^{ik_{mz}z} + E_2 e^{-ik_{mz}z}) + \frac{iE_1}{k_{mz}+k_z} e^{-ik_z z} - \frac{iE_2}{k_{mz}-k_z} e^{-ik_z z}$$
$$- \frac{iE_1}{k_{mz}-k_z} e^{i(k_{mz}-k_z)D} e^{ik_z z} + \frac{iE_2}{k_{mz}+k_z} e^{-i(k_{mz}+k_z)D} e^{ik_z z}. \tag{5.61}$$

Substituting back into (5.59) we note that the first two terms on the right hand side of Λ cancel with E_z on the left hand side. Then equating coefficients of $e^{ik_z z}$, $e^{-ik_z z}$ we find

$$E_1 = \frac{k_{mz} - k_z}{k_{mz} + k_z} e^{-2ik_{mz}D} E_2 \tag{5.62}$$

$$E_2 = \frac{2ik_z}{\omega\mu_0\sigma_f} \left[\frac{1}{k_{mz} - k_z} - \frac{k_{mz} - k_z}{(k_{mz} + k_z)^2} e^{-2ik_{mz}D} \right]^{-1} E_0. \tag{5.63}$$

Since $k_{mz} = a - ib$, where $a, b > 0$ (see Eq. (5.47)), in the limit of a semi-infinite medium $D \to \infty$ then

$$E_1 \to 0 , \qquad E_2 \to \frac{2ik_z}{\omega\mu_0\sigma_f}(k_{mz} - k_z)E_0 . \tag{5.64}$$

In the limit where $k_x = 0$, $D \to \infty$ the results of Schwartz [2, Sect. 7.1] for a semi-infinite medium and a normal incident wave are retrieved.

5.3.3 The Reflected Wave

The reflected wave is the wave field travelling downwards below $z = 0$. In this case all the slices are above the observation point, so we integrate Eq. (5.35) for δE^- throughout the material. The z dependent term is then

$$E_{fz} = -\frac{\mu_0\omega\sigma_f}{2k_z}e^{ik_z z}\int_0^D E_z(z')e^{-ik_z z'}\,dz' , \tag{5.65}$$

where the subscript f indicates the reflected component. Substituting for $E_z(z)$ from Eq. (5.45) and then integrating

$$E_{fz} = i\frac{\mu_0\omega\sigma_f}{2k_z}e^{ik_z z}\left[-\frac{E_1}{k_{mz} - k_z}e^{i(k_{mz}-k_z)D} + \frac{E_2}{k_{mz} + k_z}e^{-i(k_{mz}+k_z)D} \right.$$
$$\left. + \frac{E_1}{k_{mz} - k_z} - \frac{E_2}{k_{mz} + k_z}\right]. \tag{5.66}$$

Using Eq. (5.62) this reduces to

$$E_{fz} = i\frac{\mu_0\omega\sigma_f}{2k_z}e^{ik_z z}\left(\frac{e^{-2ik_{mz}D} - 1}{k_{mz} + k_z}\right)E_2 . \tag{5.67}$$

Hence the reflected wave is defined by

$$\mathbf{E}_f = i\frac{\mu_0\omega\sigma_f}{2k_z}e^{i(\omega t - (k_x x - k_z z))}\left(\frac{e^{-2ik_{mz}D} - 1}{k_{mz} + k_z}\right)E_2\hat{\mathbf{y}} , \tag{5.68}$$

where E_2 is given by (5.63).

5.3.4 The Transmitted Wave

The wave which emerges from the medium involves the original wave and all the waves travelling up from the slices, defined by (5.33). Hence the z component

$$E_{tz} = E_0 e^{-ik_z z} - \frac{\mu_0 \omega \sigma_f}{2k_z} e^{-ik_z z} \int_0^D E_z(z') e^{ik_z z'} \, dz' \, . \tag{5.69}$$

Evaluating the integral leads to

$$\begin{aligned} E_{tz} = E_0 e^{-ik_z z} - i \frac{\mu_0 \omega \sigma_f}{2k_z} e^{-ik_z z} &\left[-\frac{E_1}{k_{mz} + k_z} e^{i(k_{mz}+k_z)D} \right. \\ &\left. + \frac{E_2}{k_{mz} - k_z} e^{-i(k_{mz}-k_z)D} + \frac{E_1}{k_{mz} + k_z} - \frac{E_2}{k_{mz} - k_z} \right] . \end{aligned} \tag{5.70}$$

Using the definitions of E_1, E_2, Eqs. (5.62) and (5.63), we find

$$E_{tz} = -\frac{4E_0 k_z k_{mz} e^{-i(k_{mz}+k_z)D}}{(k_{mz} - k_z)(k_{mz} + k_z)^2} \left[\frac{k_{mz} - k_z}{(k_{mz} + k_z)^2} e^{-2ik_{mz}D} - \frac{1}{k_{mz} - k_z} \right]^{-1} e^{-ik_z z} . \tag{5.71}$$

The transmitted wave is therefore

$$\mathbf{E}_t = e^{i(\omega t - k_x x)} E_{tz} \, \hat{\mathbf{y}} \, . \tag{5.72}$$

5.4   Transverse Magnetic (TM) Mode

As discussed earlier, the medium is not charged but due to the inhomogeneity at the boundaries there may be a surface charge. To quantify the effect of the surface charge we start with the continuity equation

$$\nabla \cdot \mathbf{J} = \sigma_f \nabla \cdot \mathbf{E} = -\frac{\partial \rho}{\partial t} \, . \tag{5.73}$$

The time variation is of the form $\exp(i\omega t)$ hence we deduce a form for the total charge density

$$-i\omega \rho = \sigma_f \nabla \cdot \mathbf{E} \, . \tag{5.74}$$

Since the medium is homogeneous $\nabla \cdot \mathbf{E} = 0$ inside the material while at the outer surfaces the electric field has a component normal to the interfaces. To avoid confusing the surface charge (which is usually denoted by σ) with electrical conductivity we here follow [1] and denote it by Σ. From (5.74) we find the charge at a constant z surface must be

$$\Sigma(z) = \frac{i\sigma_f}{\omega} \mathbf{E} \cdot \hat{\mathbf{n}} \, , \tag{5.75}$$

where $\hat{\mathbf{n}} = \pm \hat{\mathbf{z}}$ is the normal pointing out of the upper or lower surfaces.

As stated, within the material $\rho = 0$ yet here we make the apparently contradictory statement that at any constant z surface there is a charge. However, for any given surface within the material there are corresponding surfaces just above and below, which have the opposite sign for the normal. Consequently all these theoretical surface charges cancel out. This is not the case at the outer boundaries $z = 0, D$ where there are no corresponding surfaces and so the charges can play a role there.

The electric potential follows in exactly the same manner as the derivation for the magnetic potential, which resulted in Eqs. (5.31) and (5.34), but with $\mu_0 \mathbf{J}(\mathbf{r}, t)$ replaced with $\rho(\mathbf{r}, t)/\epsilon_0$

$$\delta\phi^+(\mathbf{r}, t) = -i\frac{1}{2\epsilon_0 k_z} e^{i(\omega t - k_x x)} e^{-ik_z(z-z')} \rho(z')\delta z' \tag{5.76}$$

$$\delta\phi^-(\mathbf{r}, t) = -i\frac{1}{2\epsilon_0 k_z} e^{i(\omega t - k_x x)} e^{ik_z(z-z')} \rho(z')\delta z' . \tag{5.77}$$

The surface charge may be considered as the limit of the charge density over a thin section, $\rho(z')\delta z'$, as the section size $\delta z' \to 0$, that is

$$\lim_{\delta z' \to 0} \rho(z')\delta z' = \Sigma(z') , \tag{5.78}$$

where $\Sigma(z') = 0$ everywhere within the material, but $\Sigma(0)$, $\Sigma(D) \neq 0$.

In this case the induced electric field consists of two components. For $z' < z$ we have δE_A^+ given by Eq. (5.33) and

$$\delta\mathbf{E}_\phi^+ = -\nabla(\delta\phi^+) = \frac{1}{2\epsilon_0 k_z}(k_x\hat{\mathbf{x}} + k_z\hat{\mathbf{z}})e^{i(\omega t - k_x x)} e^{-ik_z(z-z')} \rho(z')\delta z' . \tag{5.79}$$

The induced electric field when $z' < z$ is therefore

$$\delta\mathbf{E}^+ = \left[\frac{\rho(z')}{2\epsilon_0 k_z}(k_x\hat{\mathbf{x}} + k_z\hat{\mathbf{z}}) - \frac{\mu_0\omega}{2k_z}J_z(z')\hat{\mathbf{y}}\right]e^{i(\omega t - k_x x)} e^{-ik_z(z-z')} \delta z' . \tag{5.80}$$

When $z' > z$

$$\delta\mathbf{E}^- = \left[\frac{\rho(z')}{2\epsilon_0 k_z}(k_x\hat{\mathbf{x}} - k_z\hat{\mathbf{z}}) - \frac{\mu_0\omega}{2k_z}J_z(z')\hat{\mathbf{y}}\right]e^{i(\omega t - k_x x)} e^{ik_z(z-z')} \delta z' . \tag{5.81}$$

As with the TE case we allow $\delta z' \to 0$ and then integrate. The charge density disappears within the medium leaving the surface contributions. The variation with z is described by

$$\mathbf{E}_z(z) = \mathbf{E}_{inc} - \frac{\mu_0 \omega \sigma_f}{2k_z} \left[e^{-ik_z z} \int_0^z \mathbf{E}_z(z') e^{ik_z z'} \, dz' \right.$$

$$\left. + e^{ik_z z} \int_z^D \mathbf{E}_z(z') e^{-ik_z z'} \, dz' \right] + \frac{\Sigma(0)}{2\epsilon_0 k_z} (k_x \hat{\mathbf{x}} + k_z \hat{\mathbf{z}}) e^{-ik_z z}$$

$$+ \frac{\Sigma(D)}{2\epsilon_0 k_z} (k_x \hat{\mathbf{x}} - k_z \hat{\mathbf{z}}) e^{ik_z(z-D)} . \tag{5.82}$$

On the lower surface the normal $\hat{\mathbf{n}} = (0, 0, -1)$, at the upper surface $\hat{\mathbf{n}} = (0, 0, 1)$. From the definition (5.75) we find

$$\Sigma(0) = -\frac{i\sigma_f}{\omega} \mathbf{E}_z(0) \qquad \Sigma(D) = \frac{i\sigma_f}{\omega} \mathbf{E}_z(D) . \tag{5.83}$$

Although Eq. (5.82) is more complex than that found in the previous section, Eq. (5.39), its solution is straightforward. The difference between (5.39) and (5.82) is simple and related to the surface charge terms. First we have written it in vector form since the surface charge causes the TM wave to have x, z components. Second we now have two extra terms, when compared to the TE case, which vary like $e^{\pm ik_z z}$. Differentiating both of these terms twice results in $-k_z^2$ multiplied by the original terms. This means that we obtain the vector equivalent of the ordinary differential equation (5.44)

$$\frac{d^2 \mathbf{E}_z}{dz^2} = -(k_z^2 - i\mu_0 \omega \sigma_f) \mathbf{E}_z . \tag{5.84}$$

Hence the electric field is

$$\mathbf{E}(\mathbf{r}, t) = e^{i(\omega t - k_x x)} \left(\mathbf{E}_1 e^{ik_{mz} z} + \mathbf{E}_2 e^{-ik_{mz} z} \right) , \tag{5.85}$$

where k_{mz} is the same as defined earlier. The values of $\mathbf{E}_1, \mathbf{E}_2$ follow as before by substituting into (5.82) but due to the extra terms these values will differ to those found in Sect. 5.3.2. The reflected and transmitted waves then follow in exactly the manner described in Sects. 5.3.3 and 5.3.4.

5.5 Conclusions

By describing the interaction of light with matter as light scattering from individual scatterers and then considering the effect of an ensemble of scatterers the macroscopic index of refraction and relation to the electric and magnetic susceptibilities can be derived. This approach clearly demonstrates that the reduction of the speed of light in matter is due to the scattered waves which originate from different locations in space and so, due to the finite speed of light, are phase shifted. Furthermore the wave vector component parallel to the surface is conserved and only the component perpendicular to the surface is modified by the medium.

References

1. H.M. Lai, Y.P. Lau, W.H. Wong, Understanding wave characteristics via linear superposition of retarded fields. Am. J. Phys. **70**, 173 (2002)
2. M. Schwartz, *Principles of Elctrodynamics* (Dover, Mineola, 1987), p. 234
3. R. Feynman, *The Feynman Lectures on Physics*, vol. 1, Chap. 30 (1963)
4. I.S. Gradshteyn, I.M. Ryzhik, *Tables of Integrals, Series and Products*, 7th edn. (Elsevier, Amsterdam, 2007)

Index

© The Author(s), under exclusive licence to Springer Nature Switzerland AG 2020
W. Bacsa et al., *Optics Near Surfaces and at the Nanometer Scale*,
SpringerBriefs in Physics, https://doi.org/10.1007/978-3-030-58983-7